The
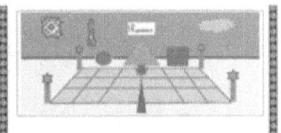

Mathematical Universe

Zeno

© 2021 Diogo de Souza
All Rights Reserved.

Contact Information:
diogodesouza7@gmail.com
diogodesouza7@hotmail.com

Zeno enters the Matrix

The trip begins with our little Zeno wandering in the hallways of his school with his heavy backpack when he met one of his teachers at the front door of a classroom. The teacher was wearing all white and had a blue skin and two earphones over his ears. The teacher gave a sign with his hands waving indicating Zeno to enter the classroom. When Zeno entered the classroom there was a very bright light that hurt his eyes. He then sat in one of the seats in front of a computer and there was a helmet on the desk. There were two other students in the classroom but soon they would be connected to the entire school district wirelessly.

The teacher said:

Lonyfaryondy: My name is Lonyfaryondy, your teacher for today. You will also be taught by Songycraype and Zack a few times. We are here about to enter a completely different world. One in which the mind and the experience becomes truly one. Please sit back and place the helmet over your heads. The helmet will automatically identify you without need of any password or username. The helmet knows its user instantly.

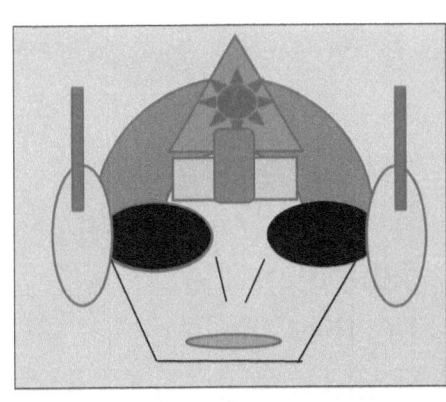

Lonyfaryondy

Zeno placed the helmet over his head and his mind then began to be transported to a complete new reality. He could see Lonyfaryondy who said:

Lonyfaryondy: Today were going to talk about electricity and some electronics. There are three components of a circuit essential for our experience here today. Besides the Battery and the wire they are the Resistor, Capacitor, and Inductor.

Current through the wires of a circuit are caused by an Electric Potential Difference (Voltage), which are an unbalance of charges that leads to their flow. When inserting a copper metal and a zinc metal in a potato, connected in series with other potatoes or with a device, electricity is allowed to flow. The positive zinc ions from the zinc metal moves out of the metal into the potato leaving two electrons that flow through the wire towards the copper metal which attracts other positive ions in the potato. The copper metal becomes positively charged and the zinc negatively charge. This process leads to a Battery and a flow of charges which can be used to turn on devices such as a light bulb.

Potato Battery

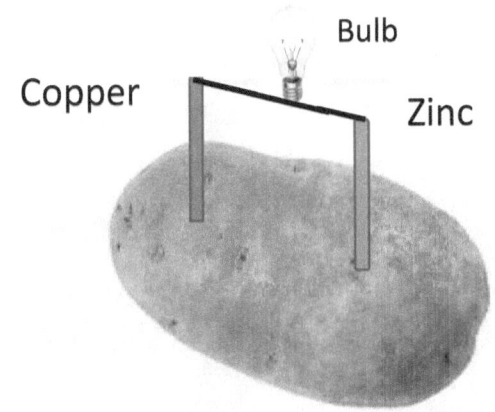

Current

Currents are flow of charges. The conventional current is understood to be the flow of positive holes which are regions in matter where Electrons are missing. In reality current is the flow of Electrons but in the 1800s it was believed that current was the flow of positive charges, and when scientists realized that it was the negatively charged Electrons that flowed they decided to keep the idea that Current is the flow of positive holes instead which is thought to be regions of missing Electrons. Conventional positive Current flows in the opposite direction of the Electron flow.

Resistor

Resistors are the parts of the Circuit which resist the flow of Current which are used to control the amount of flow of the charges per time. The wider and shorter the resistor the more charges can flow through it, while the narrower and longer the resistor the less charges can flow through it at a given time.

Capacitor

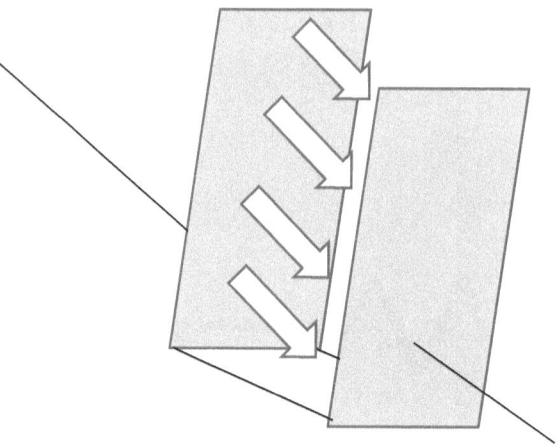

Yellow arrows show direction of Electric Field between the plates

A Capacitor is used to store a given amount of Energy and Potential Difference on each plate of opposite charge as the other. When needed, the stored Energy in the Capacitor is released to be used for a short time. The Capacitor can then be recharged for later usage.

Capacitor
Faradays

It is possible to increase the Capacitance of a Capacitor by inserting a Dielectric Material allowing a greater amount of charge to be stored per Voltage. The Dielectric Material reduces the Electric Field between the Plates and with a smaller Voltage there is an increase in Capacitance. The equation of Capacitance is Charge/Voltage. With a Dielectric Material there is less Voltage which means greater Capacitance.

Inductor

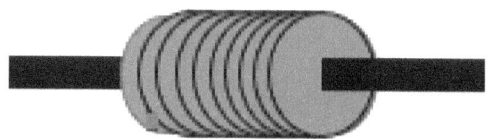

The Inductor is used to reverse Current or to generate an Alternating Current.

When inserting a magnet into an inductor a Current is generated according to the figure below.

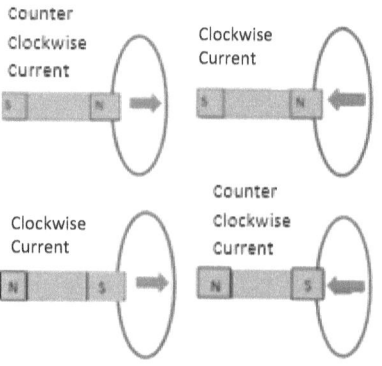

Arrows show direction of motion of magnet

The induced Current in the Inductor is always an act of nature that opposes the change in Magnetic Flux in its inside. The Current inside an Inductor also generates a Magnetic Field that opposes the Magnetic Field of the Magnet that is being inserted or removed from the entrance of the coil.

Diode

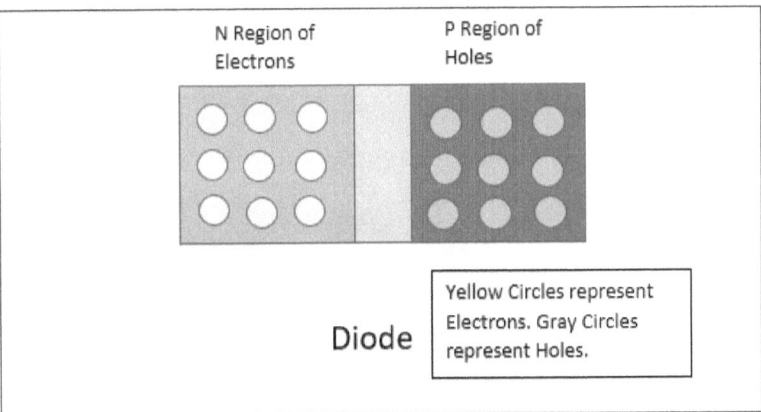

Diode

Yellow Circles represent Electrons. Gray Circles represent Holes.

A Diode is a device that only allows current in one direction. The N Region of the Diode is made of a semiconductor material with grains of Atoms of another material with an extra Electron in its Valence Orbit. These extra Electrons from these atoms gives the N region of Diode a negative charge. The P Region of the Diode is made of the same semiconductor material but this time with grains of Atoms with one less Electron in the Valence Orbit. These extra Positive Charged Holes give the P Region a positive charge.

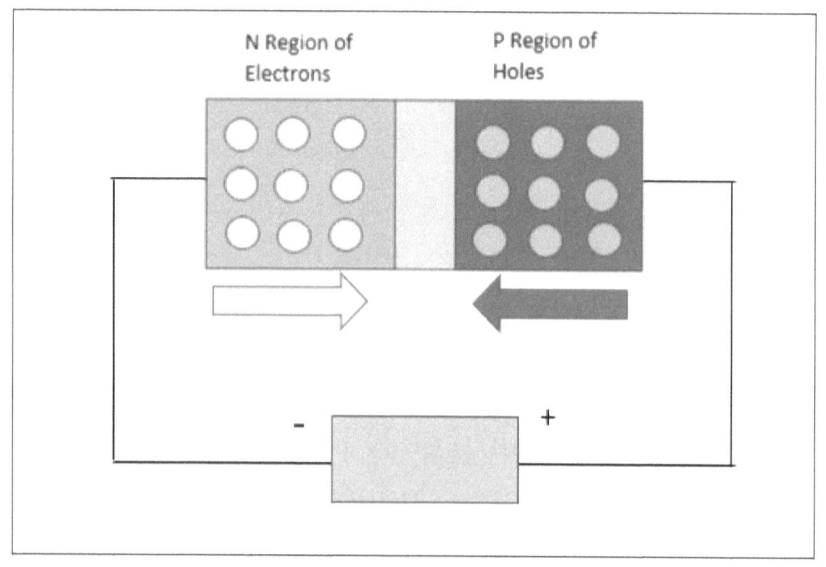

When a Battery is wired like in the previous figure the positive side of the Battery repels the Positive Holes in the Diode and Attract the Electrons which allows Current to Flow.

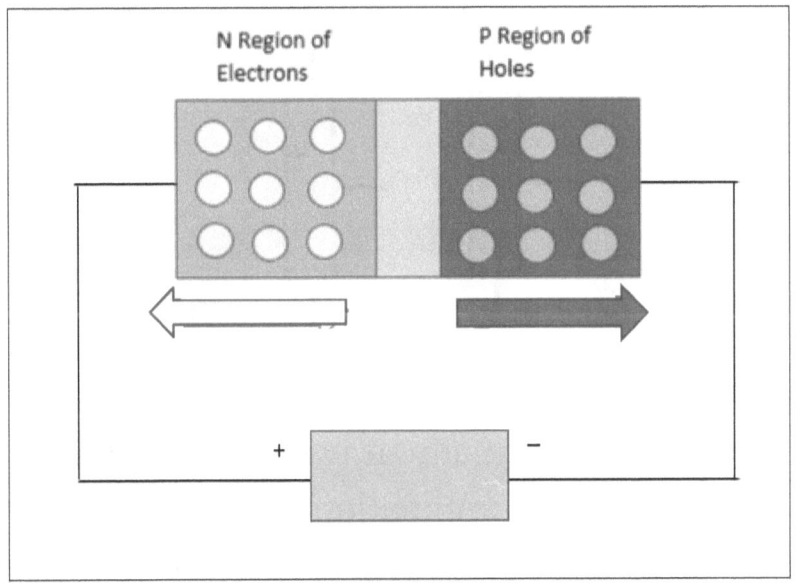

When a Battery is wired like in the figure above the Negative side of the Battery attracts the Positive Holes, and the Positive side of the Battery attracts the Electrons and this makes the charges in the Diode in being pulled in two different direction preventing the flow of Current.

Transistor

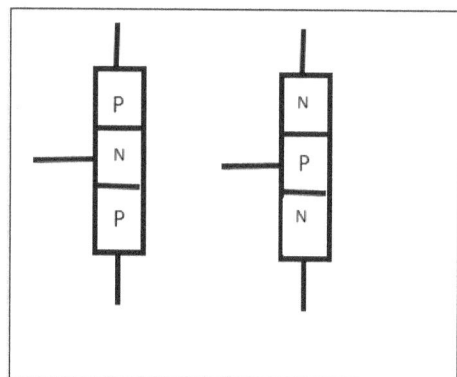

Transistors are made of two Diodes glued together. The combination inside of Transistors can be regions of npn or pnp as shown:

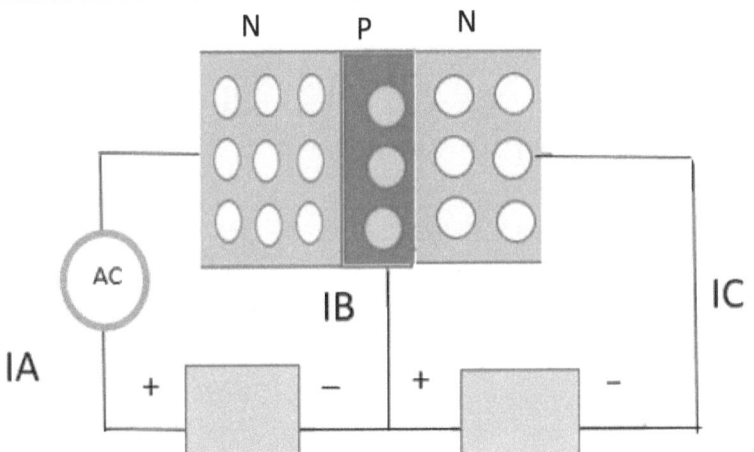

An Alternating Input Current IA pulls and pushes on the Electrons in the N Region. When Electrons are pulled the Positive Holes in the P Region are forced out into Current IB, which in effect increases the amount of Current in IC increasing the flow of charges which amplifies the Positive Signal at Current IC. When Electrons are Pushed in the N Region the Positive Holes from IB go into the P Region amplifying the loss of Current in IC. The result is an

Amplified Negative Signal in IC with a higher Amplitude.

Input Signal

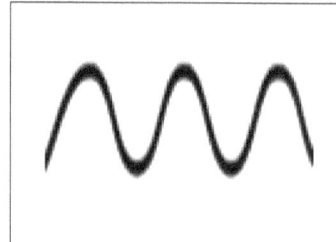

Output Signal
Higher Amplitude

Transistors can also be used as on and off switches highly useful in computer binary language of 0s and 1s, on and off signals. When there is an increase in Current it is an on signal, and when there is a decrease in Current there is an off signal. There is over a billion Transistors in the Microprocessor of a Cell Phone.

Alternating Current

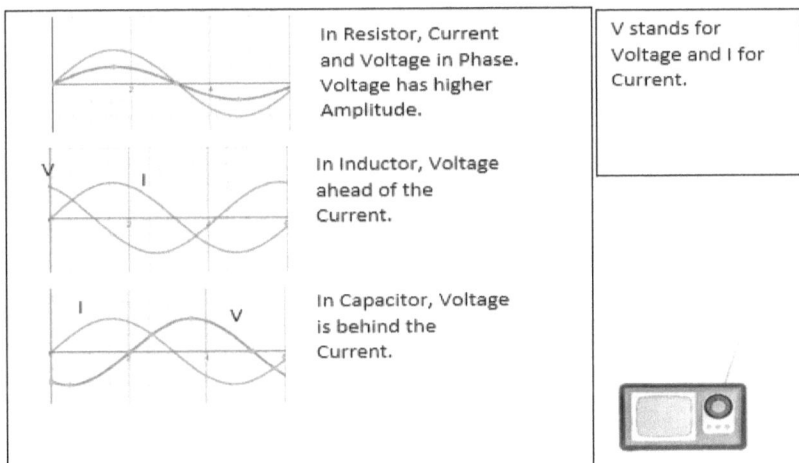

The figure above shows that a combination of an Inductor, Capacitor, and Resistor gives rise to an alternating Current in a Circuit. If an Antenna is placed in the Circuit and the Circuit is also grounded, the device becomes a Radio. The Frequency of the Alternating Current can then be changed by changing the Capacitance of the Capacitor. The

Capacitance of the Capacitor is changed when we try to move through the Radio Stations captured by the Radio. If the Frequency matches the Frequency from a Radio Station, transmission is made and you are able to hear what is being sent in the air by the Antenna of the Radio Station.

The world is surrounded by a sea of Radio Waves from all the telecommunications in the modern civilization.

Speaker

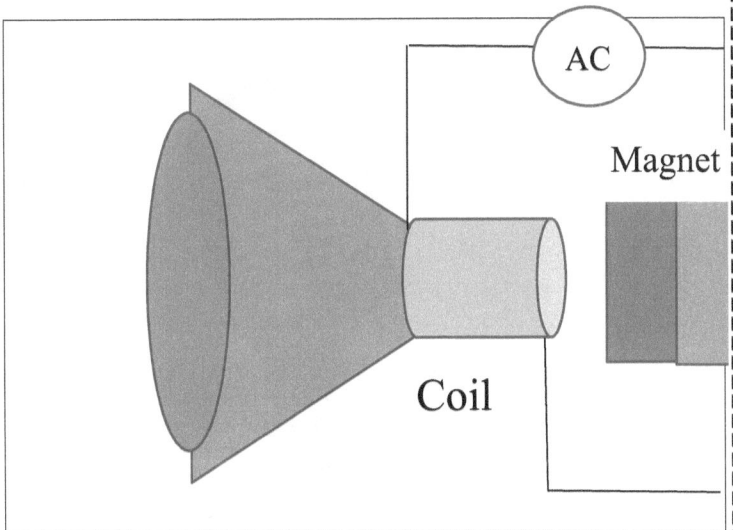

A Speaker converts an Alternating Current into Sound by forcing the Coil up and down against and towards the Magnet. This up and down motion vibrates a membrane that vibrates the particles in air leading to Sound.

Microphone

The Microphone works in reverse. Sound forces a membrane up and down, which moves the Coil above a Magnetic Field up and down, leading to an Alternating Current converting Sound into Electricity.

Cathode Ray Television

A Cathode Ray Television works using the principle shown below:

Electrons are fired from the heating filament at point C and accelerated in a vacuum tube through a Potential Difference.

Anode

Deflecting Plates

The Electrons are then deflected by Magnets or by an Electric Field to hit a specific point on the screen.

The amount of deflection of the Electrons can be adjusted by changing the field for Electrons to hit different parts of the screen.

Anode

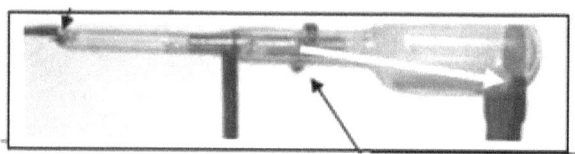

The Electrons are fired at the Phosphor Screen. Each Pixel on the screen contains a Red, Blue, and Green portion. When the Electrons hit the Green a Green dot is seen in the screen, and the same for Red and Blue. A combination of all the Pixels and the colors generate an image on the Screen.

In order to generate a complete picture, the Electrons are deflected scanning the screen from right to left and then moving downwards. The scan is done many times a second leading to an image that moves.

Flat Screen TVS use LED lights that turn on or off in each Pixel in the screen to generate the image which is made of a combination of these Pixels and colors.

Zeno was amazed by the lesson given by Lonyfaryondy when suddenly Lony seemed to move out of the scene and then Songycraype entered the Virtual World by saying:

Songycraype: Grab your sunglasses! We are going on a trip in Space Exploration.

Zeno placed the sunglasses that magically appeared on his left hand and placed them over his eyes.

Songycraype

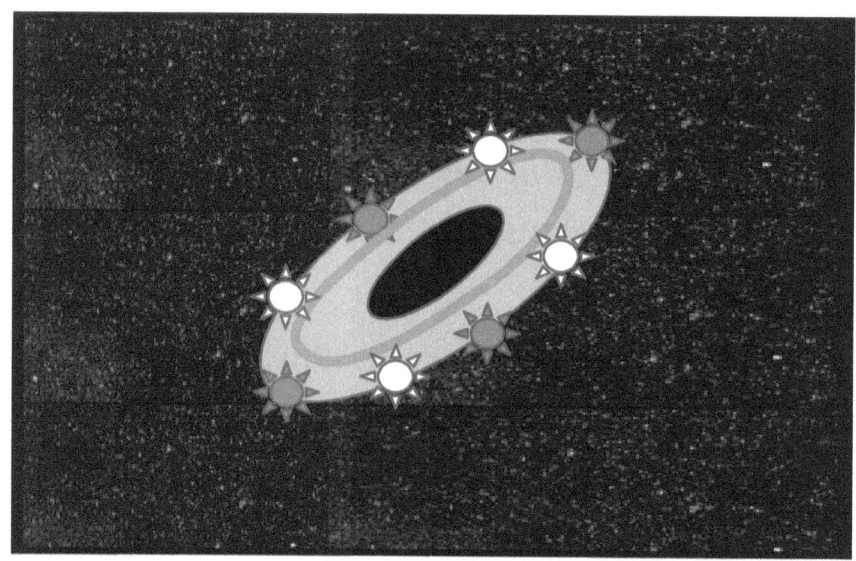

Songycraype: Since it is so difficult to land people on Mars or on the Moon, humans can start by building floating cities in space. These great cities could be in the shape of a disc in order to generate Artificial Gravity from their spin. These cities could also lead humans into space travel, roaming the cosmos towards a Planet, a Star, or staying in one location in orbit of a Planet.

Lunar bases

Before ever landing on Mars it is also essential to build Lunar Bases from which rockets can be more easily launched towards Mars without the difficult work of having to overcome an atmosphere since the Moon does not have an atmosphere. These Lunar Bases could first be built by robots and then when they are ready, humans can go and live there and help with research and new discoveries.

Fly by mars

Before landing on Mars it is important to first just simply fly by Mars without landing on it. Contemplating the Planet Mars while being in orbit of it and having a closer view of the Red World and only sending rovers to the Planet's ground to do the dangerous exploration on their own with no humans. Entire cities inside satellites in orbit of Mars can be built much before a first human landing.

Wandering cities in orbit of Mars First

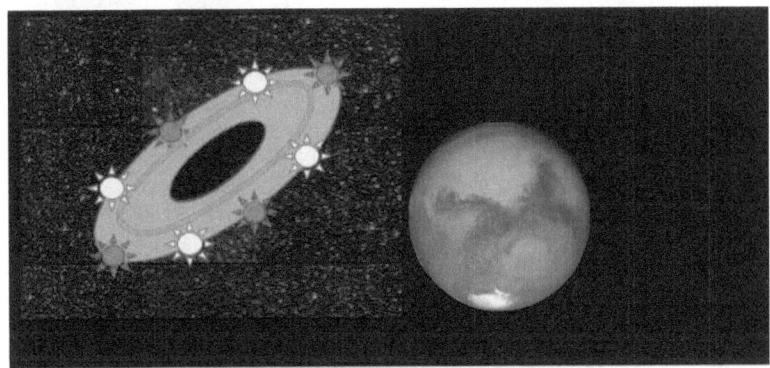

Only Rovers on the Ground

The same process can be done on Venus and any other Planet.

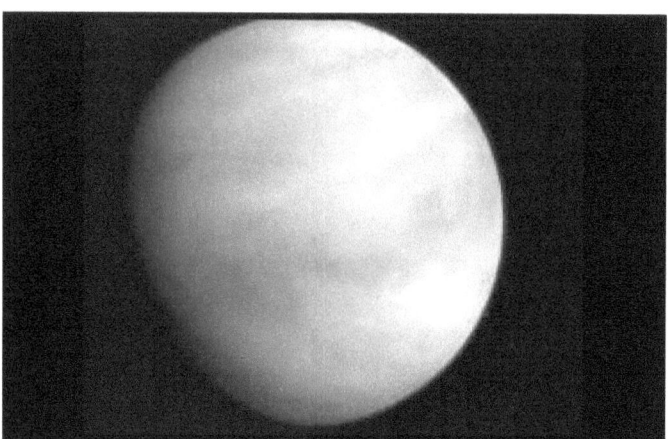

When humans find themselves ready and safe, only then they can land on the Planet that they spent so long just seeing while in orbit from their floating satellites.

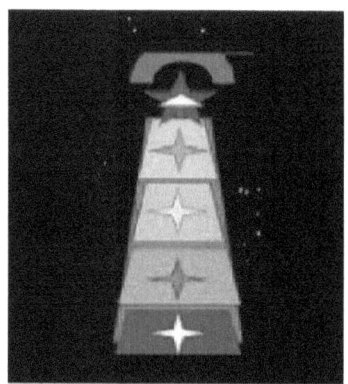

Biological Revolution

Songycraype then said:

Songycraype: That is all I have for you guys today to think about but I have an assignment to make you think. It is the following:

In this discussion you will answer the following questions with 5 sentences each, explaining your perspective on how the future of science will lead human civilization towards another revolution:

1) Explain how advances in science such as Biology could permit humans to live longer than 100 years, and how a biological process of hormones and chemicals is possibly preferred instead of electronic chip implants. Describe how all diseases may get a cure, and what would be the effect of longevity in human evolution. (At least 5 sentences)

2) Explain how a Biological Revolution could lead to a generation of humans that are smarter through a manipulation of the neurons, and chemicals that favor critical thinking in the brain. How would humans be different if the chemicals for intelligence were discovered and every human became smarter than the smarted person on earth today. (At least 5 sentences)

3) Briefly summarize the effects of longevity and increase in intelligence that could derive from a Biological Revolution in the future. (At least 5 sentences)

Songycraype then left the classroom. Zeno removed his sunglasses and Lonyfaryondy entered the room again. A bright light was turned on in the classroom and Zeno was told to remove the helmet. He could not believe on what he had just seen. For a long time he even forgot that he was in the classroom, and had his mind simply roaming around a series of thoughts and experiences to never be forgotten. That was the end of a day's trip.

Lonyfaryondy: It is over today guys and remember to do the discussion that Songycraype Hayver told you to do. Could it be possible to live 1000 years or more? Nothing seems too impossible for Science. Here goes the Biological Revolution of the Future.

Zeno contemplated the question and immediately burst into a series of ideas just like it is supposed to happen after such an intense lecture. He got up and when he was about to leave class, Lonyfaryondy gave him a packet with two other discussions and said: "Also put a thought on these two discussions written by Songycraype since they will be the next two discussions after the Biological Revolution:

1)

The universe contains 2 trillion galaxies, and each galaxy has an average of 100 to 200 billion stars, and each star is a sun. These are very big numbers which may lead one to question: "Does it even make sense based on pure reason to say that the earth is the only planet with life? If life was born by pure chance on earth, then why could it not be formed likewise elsewhere in the cosmos? The sun was formed in a cluster of stars, and over time this cluster expanded with stars moving away from each other. Today the sun is not part of a cluster and its sister stars are unknown. If the chemicals necessary for life on earth was present in our solar system, then why would it not be present in the sun's sister stars since they are formed together in a single cluster?

Did life began on earth or was brought to earth from elsewhere arriving on earth from the collision of a comet, or asteroid?

If there is life in other planets, and if this life is intelligent then why have these Aliens never contacted us? The earth exists from 4 billion years, and life began on earth 3.5 billion years ago in the form of bacteria.

Life then evolved into beings that we can see with our own eyes.

Life then moved from water to land.

Later came the Dinosaurs who lived on earth between 165 to 177 million of years ago.

Following the track of evolution the first human was born 300,000 years ago, and the first human civilization was formed 6,000 years ago.

Radio technology then only became available to us 100 years ago.

This shows that the percentage of the time in which earth's intelligent beings which are us, is capable of extraterrestrial contact with respect to the total geological time of earth is: $((100 \text{ years})/(4 \text{ billion years}))100 = 2.5 \times 10^{-6}$ %. This shows that the likelihood that an extraterrestrial civilization could enter in contact with us is extremely small. What if these extraterrestrials were alive when the earth only had dinosaurs, or bacteria, or other times earth's history, and now that we have the technology these extraterrestrials are nowhere to be found? This shows that as we humans try contact with them, these beings of other planets could also be like dinosaurs, or bacteria, or even intelligent but without radio technology. In other words, to find life in other planets is very hard. These beings could even be very developed enough to hide from us, to prevent us from ever finding them. If there are only 9 planets with life in our galaxy, and knowing that our galaxy has 200 billion stars, finding these 9 planets among hundreds of billion others is a colossal work that will take an extremely long time if possible. In other words, most intelligent beings in the universe, if they truly exist outside the earth, all have a feeling of loneliness in this silent cosmos.

In this discussion you will answer all the questions with a reply of 500 words in this text and add a conclusion which should be at least 7 sentences long:

2)

Here you will answer the following questions about wave, energy, sound, and light. You will be graded not on how accurate your answers are but on how well you explain your ideas.

1) If something is said to be small and the other to be big, it is important to have a reference. The Moon is small relative to the Earth but big relative to space shuttle. The Sun is big relative to Earth but small relative to an entire Galaxy. An atom is big relative to an Electron but small relative to a chair. Explain why is it important to have a point of reference when stating how big something is. (At least 5 sentences)

2) All matter are made of waves and all waves are formed from a combination of Sine and Cosine. Explain why is it important to have a good understanding of

Trigonometry in trying to understand the universe. The entire universe is a collection of wave functions and that gives structure to reality. (At least 5 sentences)

3) If Energy is also a wave, and Energy is conserved, explain how the universe has a limited amount of Energy since the beginning of the universe. Ever since the Big Bang the amount of Energy and matter is the same, and the only difference is that the universe expanded and this Energy and matter is more spread. Explain why Energy has to be conserved. (At least 5 sentences)

Zeno finished reading the assignments and could not wait to get home to write all three of them…..

When Zeno arrived home he received an e-mail with the following question from Zack another teacher:

Brain waves can now be transferred into electrical signals and stored in a computer that is able to match the waves to feelings in the form of images and even sound. That means that now that Artificial Intelligence is able to read people's mind by matching brain waves with thoughts, telepathy can be possible as long as people have their minds connected to each other through the Internet. The computer was taught to relate thoughts in the form of waves and to generate images and sounds that are a perfect match to what is really going on in a person's mind. Does that mean that the sense of privacy is gone? Would you connect your mind to the Internet or download your thoughts in a Pen Drive? Would you download your entire mind into Google Drive? Would you access the Internet with just a mere thought without the need to type on Google Search or speak anything, but just with a thought?

1) What is your opinion on having humans connect their mind to computers?

2) How could such scientific revolution be beneficial and harmful to people? How different would human civilization, sense of self, privacy, and overall knowledge be affected by having everyone be connected to computers and having their thoughts accessed through the internet? How would life be different when everyone is connected to each other and are able to transfer information and thoughts by using the internet with only their thoughts? (At least 200 words)

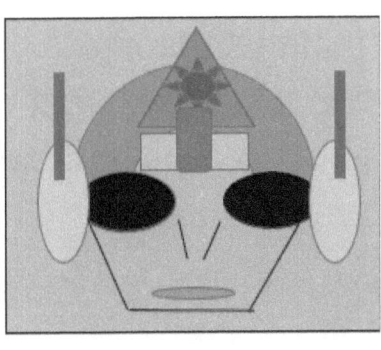
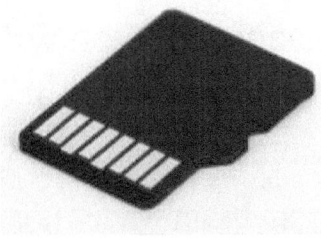

Zeno was astonished by the Revolution in Modern Science. Truly all of it seemed amazing and scary. It was clearly a golden age of thought and innovation.

The question is: Can you live without your phone or the Internet? The future is filled with possibilities and facts.

If the answer is no to the above question then that means that you are already a step ahead into a complete mind to Internet connection such as in Neuralink. People can't live without GPS and Internet anymore. Is it likely that this brain to machine fusion will happen, or will it be ethically wrong and never accepted in the society? Write down your thoughts:

From brain waves it is now possible to know what a human is doing or thinking

Zeno also found a letter in his mail box that said:

> Who will fix the problems in the world?
>
> Instead of Terra Forming Mars, how can we eliminate all deserts of the Earth and make them fertile?
>
> How can we obtain useful resources from the Moon and transport them to Earth?
>
> How can we find the cure of all diseases especially HIV?
>
> How can we eliminate hunger, poverty, by improving education and the income of all families of the world?
>
> Humans are at the very beginning of a great amount of things still in need to be done in this third millennium.

Zeno read the letter and began to think on how he can contribute for the benefit of the world and save the future.

Books of Wisdom

Zack, Zeno, and the Great City

I am Zack and I am about to start my explanation of the wonders of the cosmos to my pupil called Zeno who is sitting by a large rock. This is truly a great city which is in fact a giant computer. The pyramid is a server containing all of the knowledge of humanity from all subjects.

Each building in the shape of a cube is filled with classrooms for students to study and learn. All of the cubes are connected to the pyramid which is a server containing a lot of data. Students learn by entering a simulation in a virtual world. The way of entering the simulations is by inserting a helmet while the student sits on a comfortable chair in a classroom.

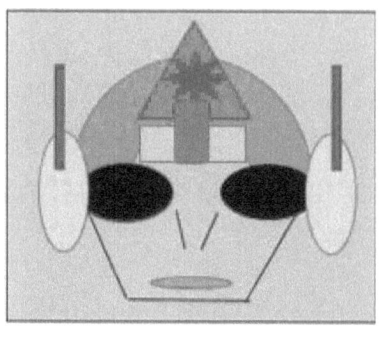

The universe is marvelous. Don't you think Zeno?

Zeno: Yes and I am ready for this adventure to better start my understanding of the reality in which we live.

Zack: The first thing to realize is looking at the four towers on the four corners of this city shaped like a square. They indicate the four corners of the world which are North, South, East, and West. Everything exists in increments. Nothing is continuous. Each direction is its own.

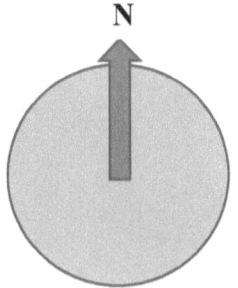

Zeno: What do you mean with continuous? What are the increments?

Zack: It is the fact that at the very small scale the great four: time, space, energy, and matter are discrete quantities which gave rise to the idea of Quantum Mechanics. Quantum means quantity. Everything is in given discrete amounts.

Zeno: Could you explain to me what discrete means?

Zack: Imagine that the minimum size of space is 2 Funs with a Fun being a unit that I am just using as an example. The length of space then can only be 0,2,4,6,8,10, 12, and so forth, but never anywhere between 0 and 2, or between 2 and 4. The discreteness here is length of 2. Now the reason why space appears smooth is due to the fact that unit is very small such as something times 10 to the negative of a big number. This makes the little pieces really extremely small thus making space appear smooth.

Zeno: So the universe is like the pixels of a TV screen?

Zack: Correct since if that was not the case humans would not be capable to understand anything. It is the fact that all things are discrete that allows measurements and predictions.

Zeno: So that means that we live in a fully rational universe?

Zack: Yes the Logos Cosmos. We are living in a reality that is well measured and calculated. Take a circle for example: Are things cyclical?

Zeno: After December comes January, and at the end of the year December again and then January in circles. A day followed by a night, and then day again. From midnight to midnight all things are cyclical. Right?

Zack: You sure understands circles. Give me more examples of circles of cycles.

Zeno: The sky is a circle or a dome. The pattern of constellations repeats every year. The seven days of every week. The four seasons, the twelve months. Life with birth, growth, reproduction, and death. Loops in a computer program. The lunar cycles. The zodiac ages. The sun around the Milky Way. Even possibly the universe itself. Breathe in and breathe out, input and outputs. Binary language with a series of 0s and 1s.

Zack: All things are either a wave or a circle.

Zeno: Please explain what you just said.

Zack: Any cycle can be represented by a wave that is a sine or cosine. Take a circle and the function cosine.

At 0 degrees the value of cosine is 1

90 degrees the value of cosine is 0

180 degrees the value of cosine is -1

270 degrees the value of cosine is 0

360 degrees the value of cosine is 1 again

And the repeats forever.

The cosine wave represents a circle:

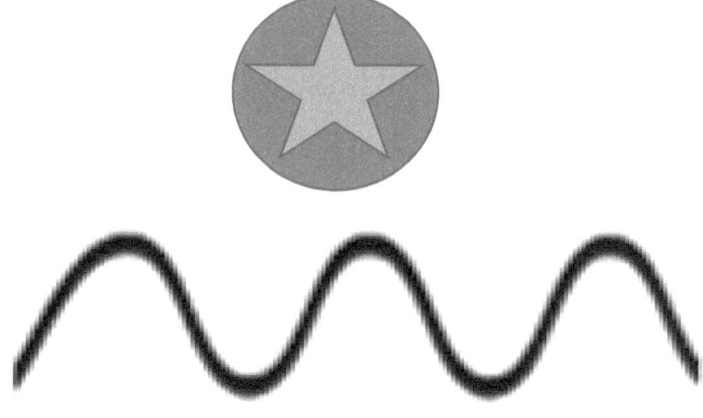

Same can be said of an oscillating pendulum. Positive amplitude at maximum displacement to the right, and negative amplitude at maximum displacement to the left.

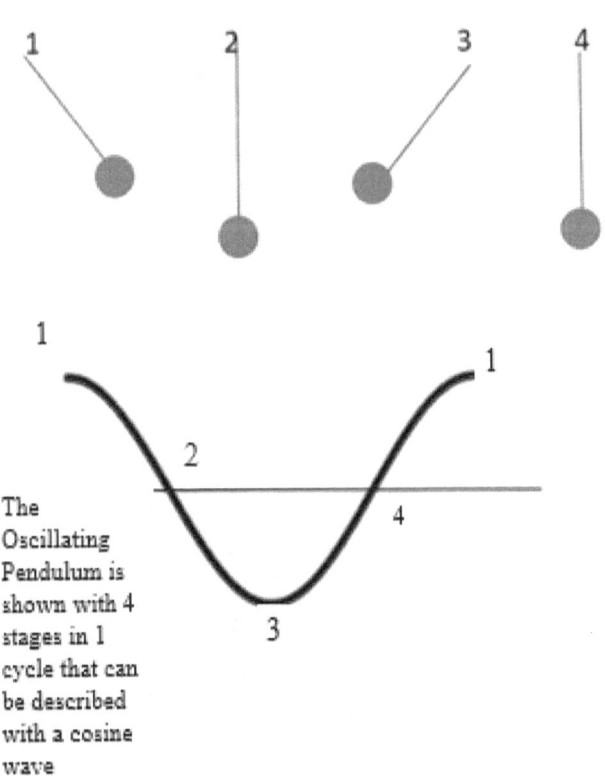

The Oscillating Pendulum is shown with 4 stages in 1 cycle that can be described with a cosine wave

A planet orbiting a star:

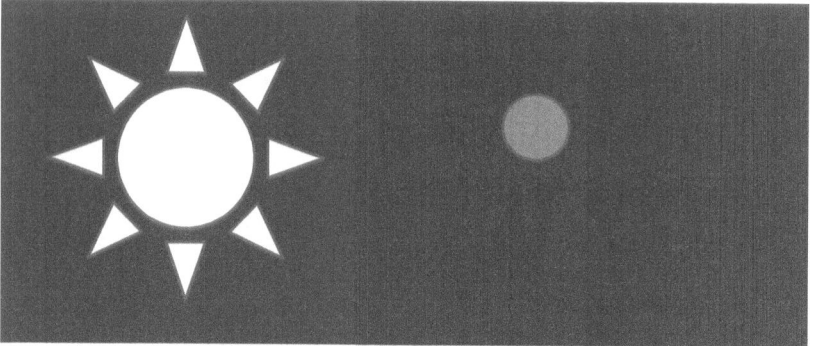

All matter is made of energy, and energy are vibration of strings which are waves with repetitions and cycles that lead to circles:

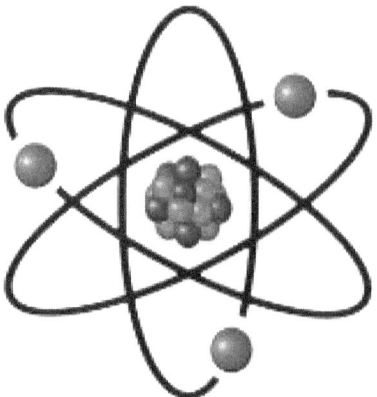

Zeno: Awesome! How can I reach ultimate knowledge of Quantum Mechanics?

Zack: Here is a ladder towards the sky. The ground state is the step on the ground, and each step is a Quantum State, and as you climb you gain more energy at increments. It is not possible to be between two steps, since the steps are discrete and separate. Climb all the way up. There is no energy above the highest energy of all, in the same way that there is no energy below the ground state.

Zeno saw a ladder reaching up to the sky:

Zeno climbed the ladder which had 35 steps. At the very top Zeno raised his arms and a wheel appeared above his head. The wheel was a portal to another dimension. Air blew Zeno up through the portal and he began crossing a Tunnel of many colors while he was spinning. It was the Xynwheel Tunnel. After crossing for about a minute he meets Vloudel the owner of this wormhole:

Zeno

Vloudel: Welcome to Xynwheel!

Zeno: What is Xynwheel?

Vloudel: It is a wormhole connecting the extremes of the cosmos. I will now open this book and explain to you about the greatest philosopher in human history: Aristotle.

Aristotle

Aristotle was one the greatest philosophers in the history of the world. He lived in the Greek Lands of the Ancient World and was possessed with a wealth of wisdom for that time in history. His writings are extraordinarily complex, and it was the standard for western scientific thinking for many centuries. When reading his books, it is possible to obtain many ideas which can be further investigated. The writings are very inspiring, and I support the fact that Aristotle's Philosophy should be taught in schools not for how updated the concepts are but for how much thinking it leads the reader into opening the horizons of the human mind to perceptions that is often unnoticed. Unfortunately, philosophy has lost importance in the western world. Not many students in universities and high schools today are presented to this great philosophical gift which is to think beyond the usual and discover many aspects of thought and perception using pure reason and observation. This is the idea that words have power, and with the skills of rhetoric it is possible to convince a massive amount of people and change the world for the better by using facts supported by reason. These can be used to demonstrate the glory of change and revolution in the world for the benefit of the society as a whole.

Source and Form

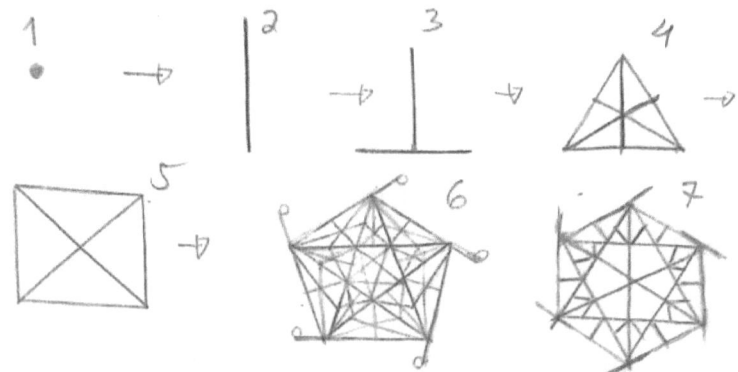

The Ancient Greeks were not aware of the existence of different atoms or molecules since these were only discovered in the last 200 years. Aristotle struggled to explain what exactly is matter and form. What makes wood a wood and a glass a glass? If they happen to have the same shape or form, then why are they still different? Why do they weigh different, or appear different? Without knowing the fact that wood and glass are composed of different atoms, the Greeks theorized on what exactly is matter and what makes it distinguishable. Aristotle could not answer this question but argued that there must be a universal source of all matter in the world but that this universal source is not a single point.

Today science acknowledges the existence of a Big Bang which is a single point which became the source of all matter and its many atoms. Aristotle, however, argued that because matter is different from other matter, then all matter could not have the same single point source since this source would only be able to produce equal matter in all the universe which is not what we see. He argued that because matter exists in many forms, a single source of all matter cannot be true. He proposed an infinite source of matter that is composed of all the infinitely possible types of matter in order to be able to form all the complexities currently seen in the world. If all matter came from the same thing, all things would be the same. If all matter instead came from one thing that is composed of all things, then all things can be derived from it.

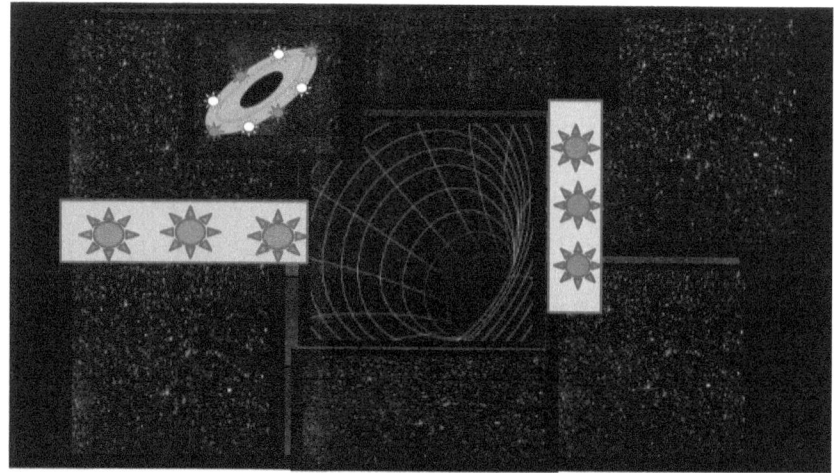

The Big Bang Theory and modern knowledge of Chemistry corrects this problem by stating that the fusion of equal atoms can produce different matter. In the Big Bang, the universe was homogenous being made of all of the exact same thing. Over time after the universe cooled, atoms were formed and currently is known that Hydrogen fuses inside stars forming Helium and other heavier atoms. Thus, it has been proven that all the matter in the universe derives from the very same source and the single point called the Big Bang. The Ancient Greeks did not know that different things could come from a same thing, which led Aristotle to state that this source was all kinds of things. In reality everything in the cosmos has a common origin at the Big Bang. Everything came from the matter and energy that the universe possessed in the beginning of time since the energy is conserved. From a single point all matter condensed to form all the atoms all the elements in the periodic table.

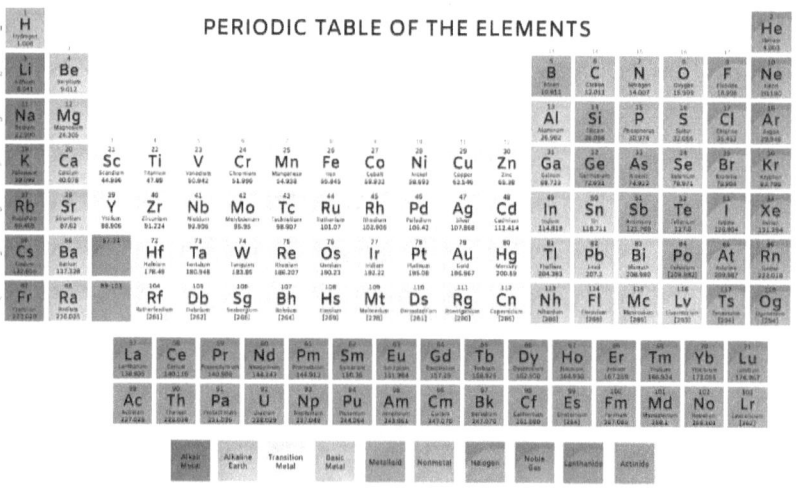

It can also be argued that a triangle or a square is made of lines but they look different. One figure has three sides while the other has four sides. Despite the fact that they are both made of the same thing which are lines, they are geometrically different.

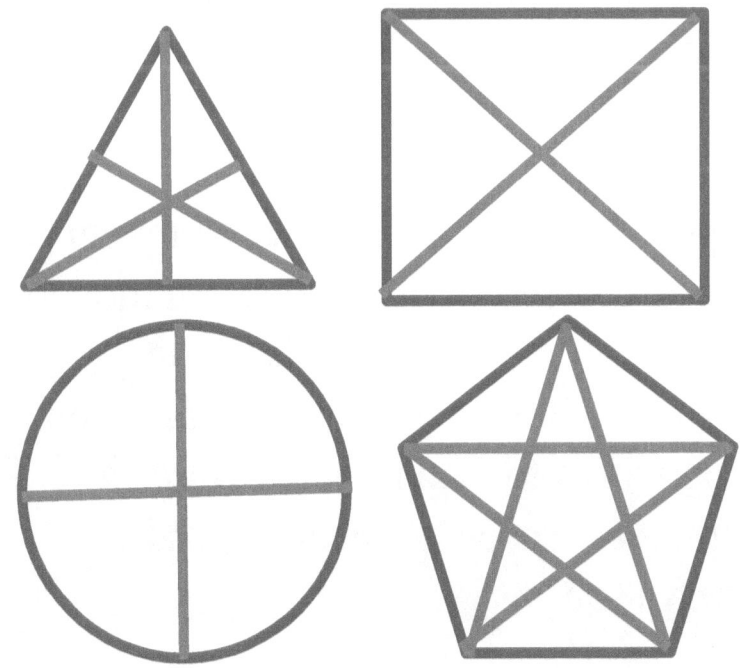

Molecules can be arranged different despite being made of the same atoms. Since the Greeks were unaware of this possibility, it can be seen that Aristotle was using the most of his effort to explain reality with the wisdom of the Ancient Greek Age. Despite of the fact that modern Science has answered the question that Aristotle attempted to solve, his effort in describing matter is quite unique. This same problem can be proposed to students, and it is a matter of curiosity to see what explanation modern students would have for this same inquiry. Philosophy is about expression, using words to lead minds towards a journey of reason, perception, and observation.

Plato argued that the universe is a reflection of a superior reality. This superior reality is the world of forms of which humans have a vague view in the world. The allegory of the cave is an example that in the world humans live an illusion of the reality while the truth is beyond this world, in this universal world of forms.

The Greek Philosophers were attempting to make sense of this world of illusion in order to explain reality which is the truth above all truths. This can be seen in how much thought was given to the world of forms and matter as seen in the works of Aristotle.

Is it possible to think of a yellow flower without seeing one with your eyes? Is it possible to see with your mind? If you thought of a yellow flower, then where is this yellow flower? It is only in your mind, or is there a superior world of forms from which the concept of all yellow flowers come from? This is the type of argument that could be made by Greek Philosophers.

Aristotle also asked a question from the fact that if A causes B and B causes C, so then A is followed by B and B is followed by C. He went a step further and asked if before A was there something that caused it, and whether there is an infinite number of causes and effects all the way to A and beyond C. This leads to the question such as what is infinity? Could the universe have a beginning at A and an end at C? Could there have been and infinite number of universes before A and will there be an infinite number of universes after C? Is cause and effect actually a circle or a cycle? Is C the cause of A, A the cause of B, B the cause of C, and C the cause of A again indefinitely? This argument is supportive of a cyclical universe that is created and destroyed, created and destroyed again indefinitely.

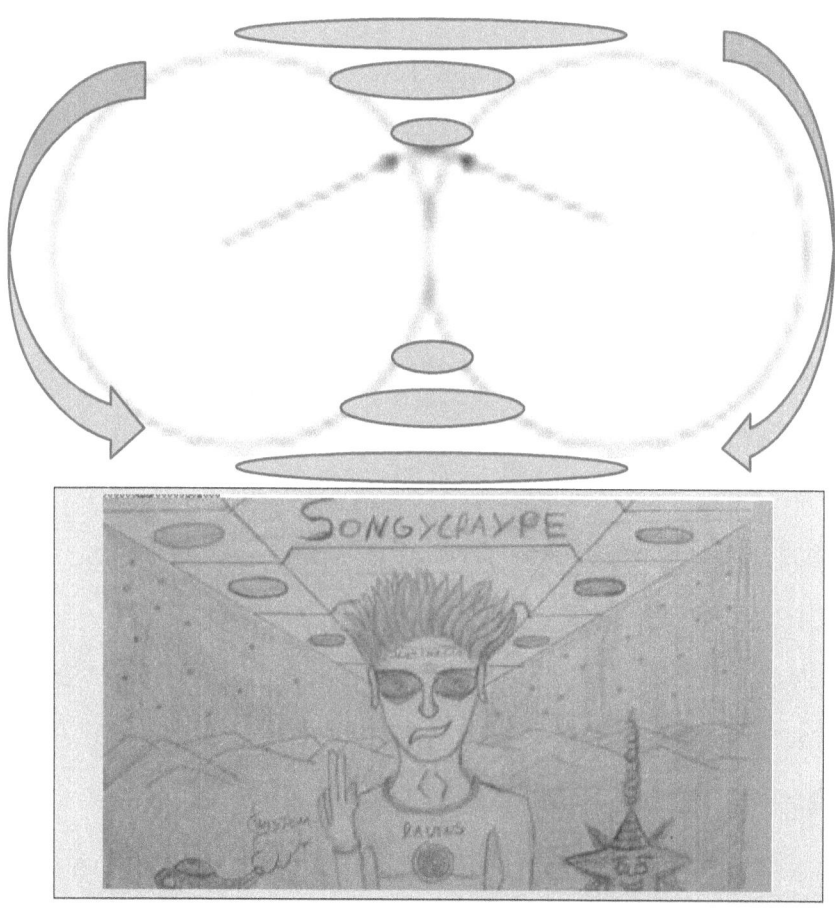

When Aristotle claimed that the source cannot be a single point, could this be true of space outside the universe? Outside the universe could exist multiple universes. In this sea of bubble universes there really is not a single point that is a source of everything, but rather several separate points, each for the origin of a unique universe. The interesting thing about philosophy is that it is possible to argue several viewpoints, and if the words are well used, it can become very persuasive. The rulers of the world are the most persuasive.

Also, part of Aristotle's thinking on infinity relates to what is between two things. If A causes B, and between A and B there is C, and there could also be D, and E then how many things could fit between A and B? He said that there could not probably be an infinite number of causes and effects between A and B. If there were an infinite number of causes, there would not be a single concrete effect. There can only be an effect from a fixed and finite number of causes. It is similar to arguing that there is an infinite number of numbers between 1 and 2 and that becomes highly complex for being unimaginable.

Quantum Physics does propose that there is the smallest unit of length, time, and energy. These units must be very small in order to appear smooth for humans and also cause us to have the illusion of it being infinite while in reality that is far from the truth. This means that there might be in nature not an infinite number of lengths between 1 and 2, but rather a finite but yet a very large number which still is not infinity but appears that way. Aristotle declared an infinity between two intervals as improbable and in the Science of Quantum Physics, infinities were also a problem which led to the concept of Planck's Length. This Planck's Length is the smallest unit of time, space, and energy possible in nature which allows the equations of Physics to better explain the observable nature of the subatomic world.

Purpose in Nature

Another common thinking in Ancient Greece is about the purpose of things.

Aristotle stated that when natural objects such a plant or an animal spontaneously have a specific trait that is not useful, that animal or plant can perish, and its species become extinct. This is very similar to the Darwin's Theory of evolution. Whenever there is an animal with a trait not able to survive as well as other animals with a better trait, the less favorable living forms with the less useful trait become extinct. Aristotle claimed that these traits occur spontaneously by pure accident, but that these accidents can serve the purpose which is survival. In the natural world, genetic mutations are constantly occurring over a long period of time. These mutations become noticeable in species after millions of years. Religion may claim that there is no accident in nature, but just the fact that there have been so many extinctions of animals due to them having less favorable traits makes one wonder why would nature do extinctions on purpose. Unless the natural world is not very benevolent but is a fallen creation such as mentioned in some scriptures.

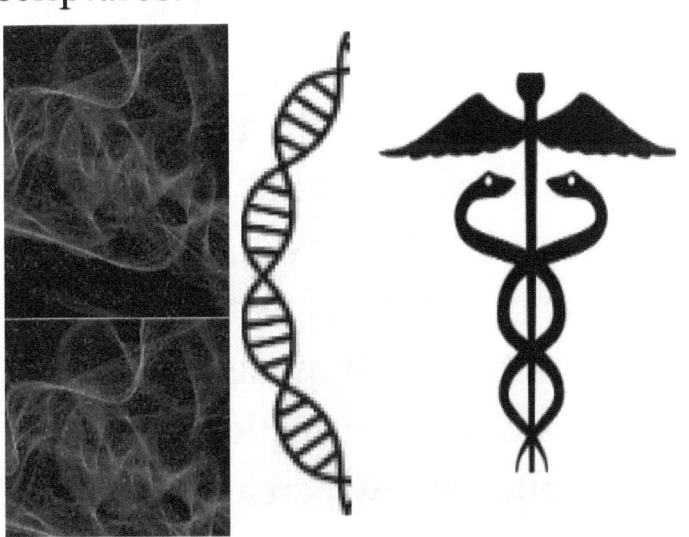

According to Aristotle, these changes in the natural world are from pure chance and these blind transformations may benefit some animals and cause the extinction of others. The reality is that the natural world is indeed very harsh. Civilized human beings can easily see that the survival of the fittest do not and cannot be applied in the human society. Humans could in fact change the rules of the natural world, and possibly build a more just reality than the one currently seen. Mankind unlike the other animals in nature have reason, and if there is something that appears not fair in the natural world, humans who are capable of making modifications in nature unlike any other known creature can indeed change what is not identified as just. Knowledge is power and all humans need to be united into a one civilization of thinkers. Only humans have the intelligence to understand justice, and we have the power to revolutionize everything around us in the name of justice. That could be recognized as the purpose of being intelligent, be it accidental or in fact developed through a series of mutations that had the intention of forming humans.

This brings to question whether the natural world has characteristics that have a purpose or are these traits accidently used for a purpose. Is the natural world truly accidental? There is a rich complexity of natural events and phenomena that brings order to chaos. It is really interesting if the chaos were able to arrange itself so well as to form a world that is rich in details and even favorable to form life and humans. These are the greatest mysteries of our existence.

Humans can be the source of a new movement and idea that will change the natural world

Thinking about the Soul with the help of Aristotle

Aristotle also thought about the nature of the soul. What is the soul? Is the self the entire body, or is it just the head? Can there be a consciousness without a physical organism? What exactly moves the limbs of an individual? Is the concept of I or self fully formed from the actual brain, or is there truly a spirit or a soul that uses the body for actions? These are questions that are not exactly what Aristotle asked but which can be derived from his research on this topic.

As can be seen, Aristotle did not know what is the true nature of the soul. He questioned whether plants have soul since they grow and also move and whether air or land also have soul. He wondered if there is a soul in everything or if it is only a characteristic possessed by animals. He considered humans a type of animal with the ability to reason and ask these questions. He argued that a soul may not be made of parts but made of a single thing the main cause of movement of the bodies. This leads one to ask if plants have soul, then where is it? If air has soul, what kind of soul is it? Is it conscious?

These questions are not very different from Eastern Philosophy which claim that all the universe has a single soul. It is not clear if Aristotle was influenced by Eastern Philosophy based on the questions that he asked. Even if there is a soul in everything, what exactly is it? Humanity does not have an answer for these questions yet.

Is the natural world completely blind or does it possess a spirit or a soul guiding it? After physical death what happens to the soul, and where does it go? Is our consciousness part of physical body, or does it exist beyond the material cosmos? Is the soul really made of one part or is indeed very complex? There are unending questions regarding the soul in which it is nearly impossible for humans to answer with 100% certainty since it is difficult to test it in an experiment. We are only left with philosophy and hypothesis.

Aristotle appeared to believe that the soul is where we are, and the human body is what the soul is using. We are riding in the body for a temporary existence in the world. He also believed that once the body dies, the soul moves to some other place, although he was not clear where does it go. In Eastern Traditions, it is claimed that the soul leaves a dead body and incarnates in another one. Aristotle does not seem to mention religion, and he investigates the nature of the soul as a scientist. He constantly states his questionings, unable to know exactly what is that makes people and animals other than the body that we use in life. It can be concluded that the soul is the eye of the mind and we are using our body for a brief amount of time. Aristotle seems to indicate a belief in after life, and the eternity of souls through several ages but he avoids talking about religion and spirituality in general.

Perception:

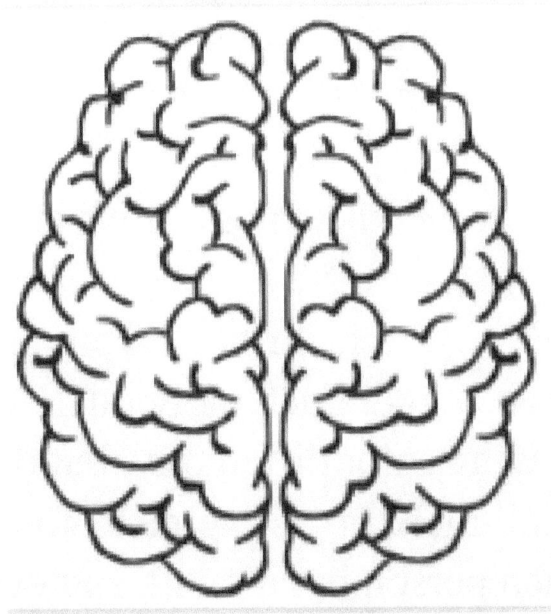

When a person sees something with the color red, it appears red because of the color that it has, but the way a specific individual perceives the color red may differ from other people. There is a difference between being red and appearing red, from what is perceived as red, or related to the color red. A person might make a connection of the color red with blood and dislike that color because of that similarity. Other people may perceive red as the color of love and charm and may have a preference for that color. Being red is different from being perceived as red. Each individual may perceive the color differently, but they should all more or less agree that it is red. Each person has a mind, and eyes.

The senses are what transmits the information that is external to the mind which is internal. Obviously if the senses can't see the color red, the mind will simply not be able to distinguish it. In other words, there is a large difference between being red and being perceived as red. Each person will see the object from a different angle, exposed from a different light, and will possibly argue specific characteristics similar or different from other people. This leads to the conclusion that each person is in the center of an individual imaginary universe being perceived differently from others. No two souls are exactly alike and will perceive things differently.

Aristotle Part 2

Perception:

Aristotle explained in his writings about the senses. The combination of an individual's senses creates the reality that he or she lives. The combination of sight, smell, heat, and touch allows us to know what is happening outside of us and molds how we feel in our minds. There is also the art to distinguish what is felt such as understanding that not all yellow things are the same object. A flower can be yellow, but a rock can also. A rock is not a flower. Not all things that are the same color are the same thing. We need a mind to perceive the difference in being able to tell one thing from another. You can only truly see if your mind is conscious of what is being seen by the eye. Without the mind the eye is blind. Viewing an object and remembering it in the future is also another aspect of the senses. Where is the memory of the object kept in the mind? Does the object still exist in the mind? Is the object and the sense that sees or remembers the object the same but yet different? The senses are always ready to experience something new. The view of a different flower, the sight of a peculiar sunrise, the feeling of a breeze. The senses are potential to experience something. There are the things that act and the others that are acted upon. This is the duality of a sensible object and the sense that is able to perceive it. Sensing and the object that is perceived in an individual's mind could be the same one thing, but in essence they are indeed very different. White and black would not exist if everyone was blind. There would be no flavor if everyone had no sense of taste. The senses and the object being perceive are one, but still different. For Aristotle music are harmony and ratios. We can conclude that since sound is made of ratios such as wavelength and frequency, then hearing is also.

Aristotle said that taste and flavors could also be ratios bitter or sweet, and the colors have degrees of brightness or darkness. Could beauty also exist in ratios, or fine proportions? Sensations then are ratios also and excesses lead to pain and destroys it. What makes perception of one thing different from the other? It is how we sense things that leads to our concept of differences, but what leads to the recognition of difference is one's mind. That is the mind of who is observing. The thoughts are divided in essence but not in space since the mind is one. Today in Quantum Physics it has been proven that with just the act of observing a Particle an individual changes that Particle. That means that the act of observing can change the entire universe. Everything is interconnected and entangled.

Zeno: Awesome! Do you know anything about Quantum Entanglement?

Vloudel: According to the Pauli Exclusion Principle no two Fermion Particles can occupy the same Quantum State. That means that if you reverse the spin of an Electron in your hand, you immediately reverse the spin of an Electron in say the farthest galaxy from Earth. This is what we call entanglement. One change made here affects instantly the farthest place in the universe. All things are entangled. The universe is big but entanglement brings to mind that it is in fact small. The changes move faster than the speed of light. They are instantaneous.

Zeno: Thinking about instantaneous reminded me of time. What about time? Is it possible to travel in time?

Vloudel: Yes. Indeed there are particles that travel to the past, and we are travelling towards the future. When a particle moves from one location to the other, it travels through all possible paths between the two points. When we attempt to track the particle we find only one path, but without observing the particle it takes a superposition of several paths. Quantum Physics has shown that just by the act of observing the universe changes. Do you act the same way when no person is looking at you?

Zeno: No!

Vloudel: Particles exist with a sea of probabilities and wave functions but when you see it, this condition collapses into a single state. What was the particle doing before you observed it? Possibly having a trip to other parallel universes, and even travelling back and forward in time.

Zeno: It is crazy and mind bending. Are there particles in me that came from another dimension?

Vloudel: There are particles in your body right now that came from the future, and others that came from the past, and others that came from a parallel universe. We live in a multiverse. It is really crazy like you said.

Vloudel opened another book and began reading:

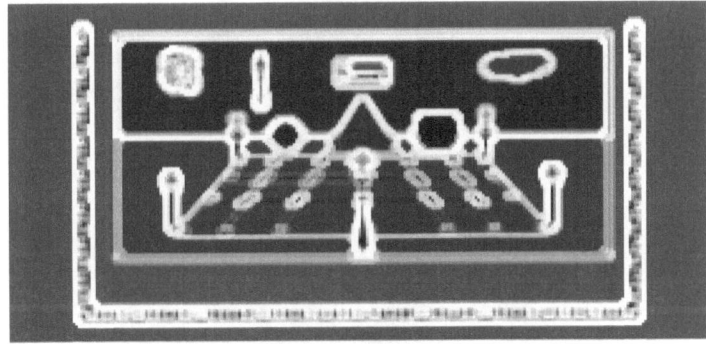

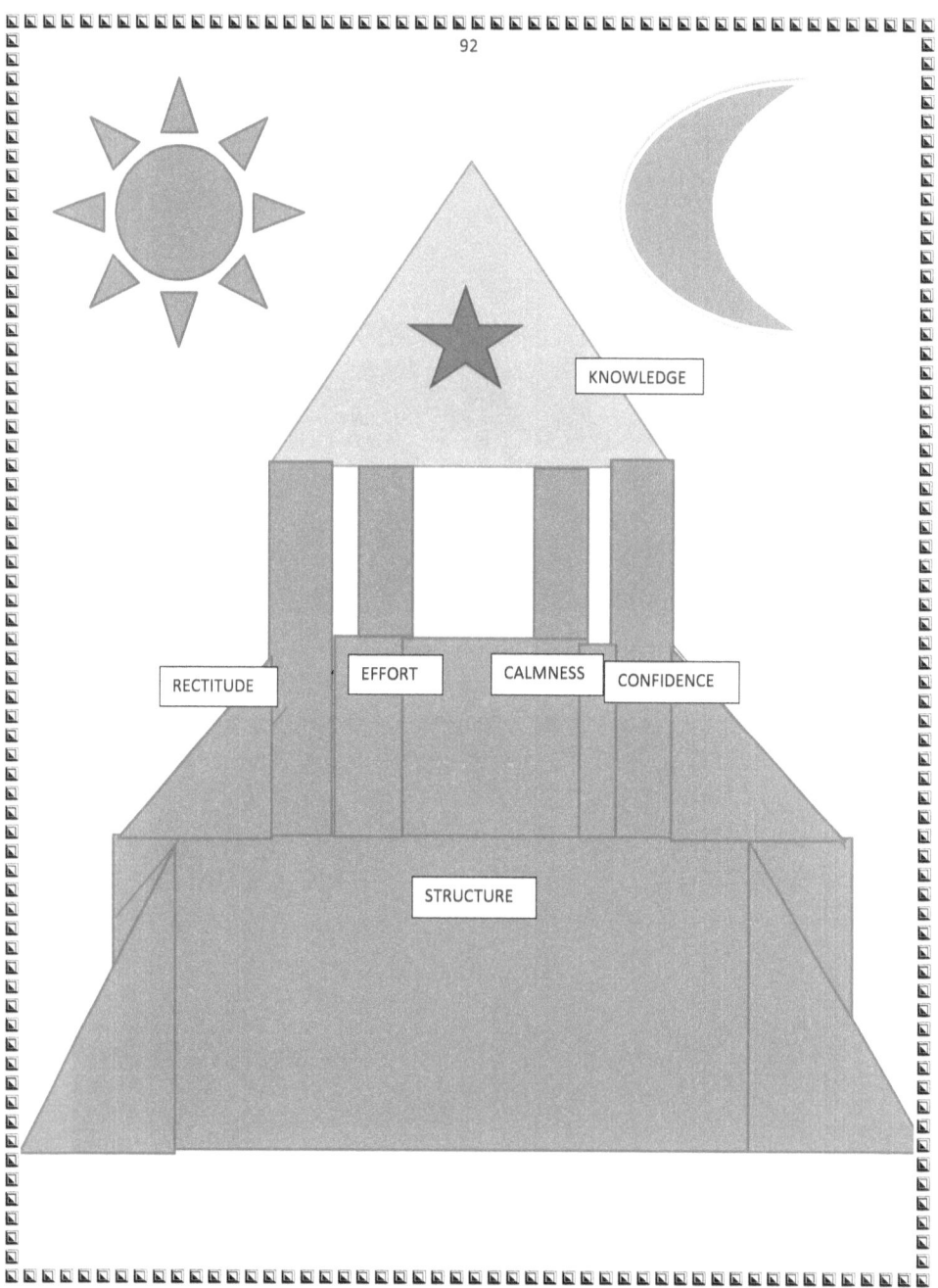

The Cosmos and the Superposition of Waves:

All matter and energy in the universe exist in Hilbert Space. This space is filled with waves and a superposition of these waves is added together forming the structure of the cosmos. It can be reasonably stated that the cosmos vibrates since particles exist in clouds of probability waves in an abstract manner in which Physicists still do not know what that exactly means. Each wave within the cosmos can be made through a combination of other waves similar to having numbers 5,6, and 7 as the basis from which their sum can be 5+6 = 11, 5+7 = 12, 6+7 =13, or 5+6+7 = 18. The numbers 11, 12, 13, and 18 were formed from the sum of numbers which in this case serves as an example of the basis functions in Hilbert Space. All matter seen in the cosmos is composed of the sum of waves in the quantum space.

Energy Levels and Standing Waves

The energy levels of a particle inside an infinite square well potential generate a wave which appears like waves in the string of guitar. Waves in strings are called standing waves, and the fact that particles behave similarly inside an infinite potential well is an example of how the Schrodinger's Equation can be used to map this phenomenon. In reality there is no infinite square well potential in the universe. All energies are limited and there is always a chance that particles can penetrate the barrier. By imagining an infinite square well potential, physicists are exploring the concepts on how particles would behave when they were found inside such a potential. From the wave function of these particles, the absolute value of the coefficient squared can be used to find the probability of the particle at being in that energetic state, and also explore the fact that wave functions are a sum of other wave functions in Hilbert Space.

Energy Levels and Increments

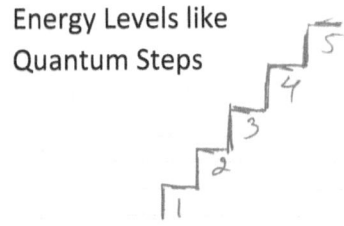

Energy Levels like Quantum Steps

Fig: Energy Levels are light steps of a stairs. You can only place your foot on the steps but not anywhere in between. The universe only allows discrete levels of energy, angular momentum, spin, space, time, and so forth.

Uncertainty Principle

By knowing the momentum of a particle really well the wave function becomes wide making its position less certain. By knowing the position of a particle really well the wave function becomes more local making its momentum less certain due to it being difficult to read the wavelength inside the wave packet.

Spin up

Spin down

Superposition of two states explained:

In Quantum Mechanics it is learned that particles can have a superposition of two states with up and down spin together at the same time. That does not mean that the particle is spinning in both directions but instead that it exists in an abstract condition that allows the mathematics to work when describing its nature.

What is Spin and Angular Momentum of a Particle?

Quantum numbers may not have a real and concrete meaning for the physical world. When physicists say that a particle is in a quantum state, they mean that the particle fits within a pattern seen at the subatomic level. Particles may not spin or have an angular momentum like macroscopic objects. First of all, particles are not physical objects. Physicists had to call a particular quantum state with a name, and they chose angular momentum and spin. A particle has never been seen spinning or orbiting a nucleus. Particles exist instead in a wave of probabilities within a very abstract reality beyond physical comprehension. No two particles that are Fermions can occupy the same quantum state, and that is why a change in one region of space automatically changes another to obey the Pauli Exclusion Principle. It states that particles exist in a unique quantum state. That reasoning is what brought scientists to confirm the entanglement of particles.

Vloudel: Now I have read to you many important things from this strange dimension. I think that it is time for you to return to the desert.

As soon as Vloudel said that, Zeno once again entered a colorful tunnel the Xynwheel Wormhole through several dimensions. He was spinning as he moved forward floating and travelling very fast. At the end he landed on top of the ladder at the highest step. He came down all 35 steps towards the ground and met Zack.

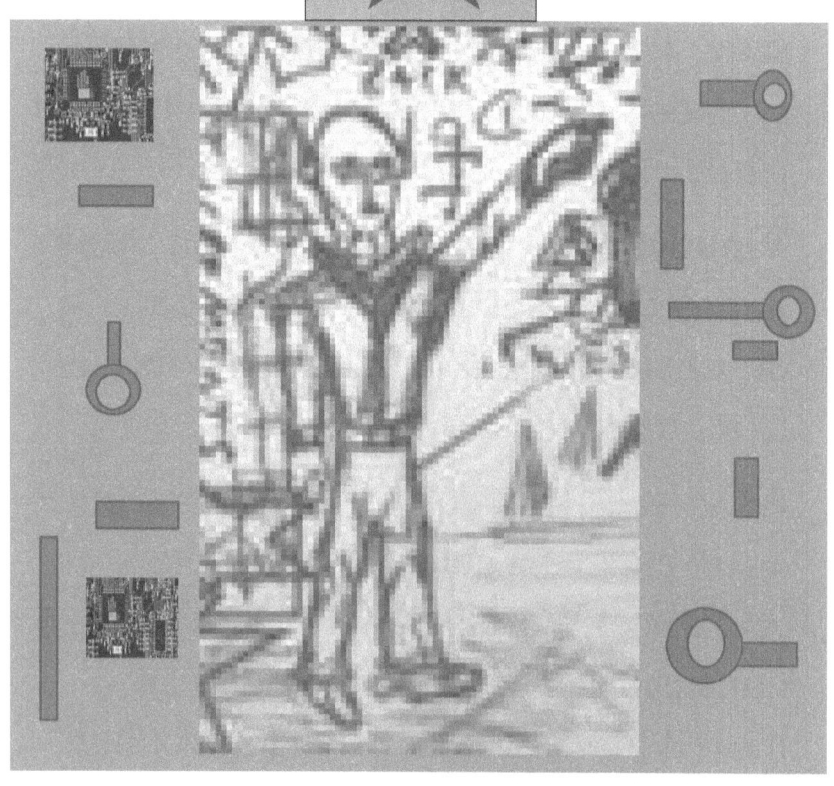

Zack: Oh I am sorry. This whole trip was a simulation. Let me remove it for you.

The helmet was removed, and Zeno found himself in a classroom seated in front of a computer with other students.

Zack was his teacher.

Zack: Ok guys, class is over, see you all tomorrow.

And that was the end of a day's lesson.

Ogoyde Azuous

© 2021 Diogo de Souza
All Rights Reserved.

Contact Information:
diogodesouza7@gmail.com
diogodesouza7@hotmail.com

This story begins with our little Zeno walking his camel over the Sahara desert. He pulled the animal gently with a rope while walking on foot. The camel was his best friend in the desert. He loves the desert and contemplates the blow of the wind under the bright sun, and the slow moving grains on the surface of the yellow dunes. All he could see until the eye meets the horizon was only him, the camel, and many of his bags hanging on it. When alone in the desert that is when his imagination flourish, and he begins to think of many things about his life, past, present, and even the future. After travelling for about an hour with no sign of civilization near as usual a strange thing happened. A crazy looking figure shows up in front of him from a lightning.

It was Vloudel coming from one of his trips through Xynwheel:

Zeno says: Oh not again. Here comes Vloudel. Is it time for a lesson? Tell me what you brought this time.

Vloudel: Can you live without energy?

Zeno: Here we go! I bet there is something interesting in the question. No, we can't live without energy. Here goes the craziness.

Vloudel: What is it?

Zeno: Energy is something that excites matter maybe. You are really into it today.

Vloudel: Can you live without it?

Zeno: You need it to survive. Energy is power but slow down Vloudel please!

Vloudel: Can there be anything without it?

Zeno: We need something to live and even exist, and that is called energy which is life.

Vloudel: Yes. All things came from energy according to Einstein's famous equation $E = mc^2$. Energy can be converted to mass and mass can be converted to energy. All things are energy and energy is light.

Zeno: What do you mean when you say that energy is light?

Vloudel: Quantum Fluctuations occur all the time throughout space. Light becomes a particle and an antiparticle briefly and when the two meet again the mass becomes light which is energy. The entire universe might have been created by such fluctuation and all we see around us is this brief existence before annihilation.

Zeno: That is scary, but it does not seem brief.

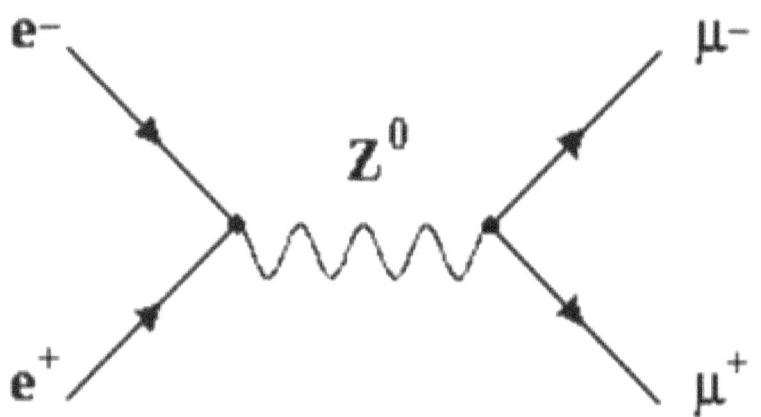

Vloudel: When you travel at the speed of light time stops. That means that for a photon it has been less than a second since the Big bang. That sounds brief to me since time is relative and it flows different for different parts of the universe.

Zeno: Wow!

Vloudel: Come with me across this portal. I need to introduce you to the cosmos.

Zeno was then pulled by Vloudel through the portal, and when he crossed it he was introduced to space with many stars all around. They were both floating in the void.

Zeno: How beautiful. There are just too many stars. It seems that the sky is filled with spots like a disease.

Vloudel: Not disease but rather light.

Zeno: Wow! Bright light everywhere. I might need sunglasses.

Vloudel: Here it goes. I got one for me too.

Vloudel then gave the sunglasses to Zeno, and he also placed one over his eyes since the radiation from the stars was very strong.

Vloudel: Light is an Electromagnetic Field. It is composed of an Electric and a Magnetic Field that oscillates 90 degrees from each other intertwined.

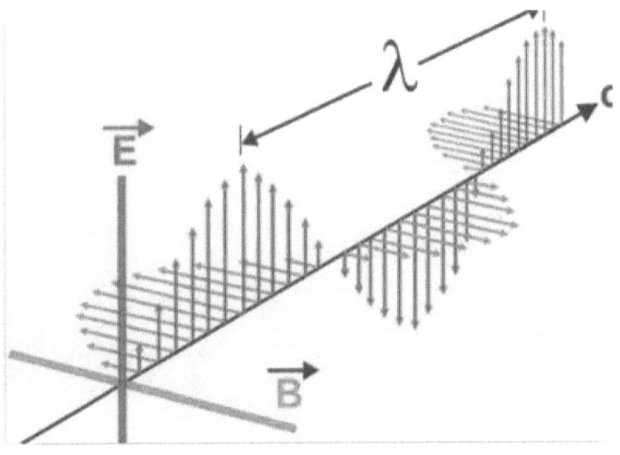

Since they are intertwined one change in one causes an immediate change on the other. That can be shown in Maxwell's Equations. Here is a tablet from Earth.

Vloudel then showed Zeno a tablet with Maxwell's Equations:

$$\nabla \cdot E = \frac{\rho}{\epsilon_0}$$ Divergence of Electric Field

$$\nabla \cdot B = 0$$ Magnetic Fields do not diverge

$$\nabla \times E = \frac{-\partial B}{\partial t}$$ Change in Electric leads to Magnetic Field

$$\nabla \times B = \mu_0 J + \frac{1}{c^2}\frac{\partial E}{\partial t}$$ and Vice Versa

We are all made of energy, and light is a form of energy. There is a great spectrum of light and what leads to its different forms is a change in frequency and wavelength. They are all light but with different frequencies. Light is everywhere so that means that all things are just frequencies:

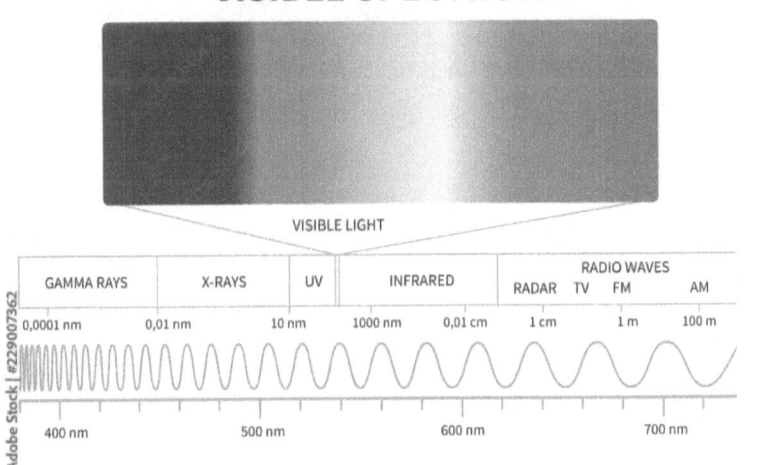

The most energetic of them all is Gamma Rays which are created in interstellar and intergalactic space.

Then comes X-rays useful in medicine to take pictures of bones in the human body.

Ultraviolet which gives you that tan from the sun.

Visible light composed of the seven colors of the rainbow which in order from most energetic to least:

Violet
Indigo
Blue
Green
Yellow
Orange
Red

Next is Infrared which is heat. Yes the heat you feel is a form of light.

Then comes Microwaves which becomes trapped heat in microwave ovens which heats your food.

Last is Radio Waves, which are useful in telecommunications, wifi, and so forth.

They all share the same nature and that is of being a form of light or energy which travels at $3.00 \times 10^8 m/s$ in a vacuum. What makes each of them different is their frequency and wavelength.

Frequency times wavelength is equal to its speed which is constant in a vacuum, but slows down through different mediums which are not a vacuum.

Zeno: Wow! That is a lot of information in one single spoon. Vloudel then continued by opening and reading one of the books written by OGO from a mysterious bag:

Electromagnetic Fields and Forces

An Introduction to Maxwell's Equations:

Let us think of a simple magnet that is a dipole never a monopole. It has a N, North Pole, and S, South Pole. The magnetic field points away from the North and towards the South Pole.

Inserting a magnet with its N side into a loop will cause a current counterclockwise.

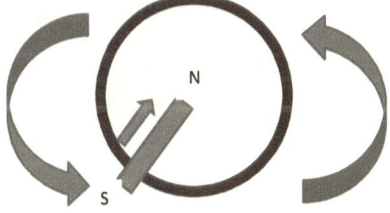

Using the same magnet and moving away from the loop will cause a current clockwise.

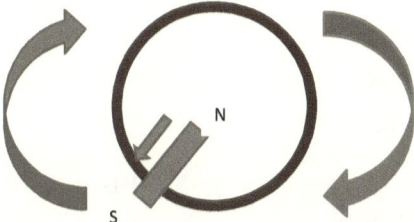

That is because nature abhors change.

In both cases there is a change in magnetic flux. The current on the wire tries to preserve the magnetic field that was before.

In a loop, a current clockwise will cause a magnetic field into the page.

In the same loop, a current counterclockwise will cause a magnetic field out of the page.

Inserting a magnet inside a loop with its S side will cause a current clockwise.

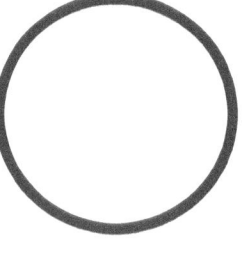

Draw what is going on according to the statement above.

Vloudel told Zeno to draw on the circle on this page and on the next what would happen by inserting or removing the magnet with its south pole facing the coil.

You the reader of this book can also draw with Zeno.

Moving the same magnet away from the loop will cause a counterclockwise current.

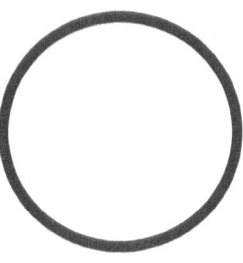

Draw what is going on according to the statement above.

A change in magnetic field causes an electric field.

Similarly, a change in an electric field will cause a magnetic field.

An electric field diverges from a source.

A magnetic field curls around itself, and its divergence is zero.

There are no magnetic monopoles. A magnet will always have a North side followed by a South side.

An electric field can be monopoles such as a single negative particle.

An Introduction to Maxwell's Equations:

Maxwell's Equations are able to explain all of Electrodynamics. I start this book with a quick summary.

Let us think of a simple magnet that is a dipole never a monopole. It has a N, North Pole, and S, South Pole. The magnetic field points away from the North and towards the South Pole.

Inserting a magnet with its N side into a loop will cause a current counterclockwise.

Using the same magnet and moving away from the loop will cause a current clockwise.

That is because nature abhors change.

In both cases there is a change in magnetic flux. The current on the wire tries to preserve the magnetic field that was before.

In a loop, a current clockwise will cause a magnetic field into the page.

In the same loop, a current counterclockwise will cause a magnetic field out of the page.

Inserting a magnet inside a loop with its S side will cause a current clockwise.

Moving the same magnet away from the loop will cause a counterclockwise current.

A change in magnetic field causes an electric field.

Similarly, a change in an electric field will cause a magnetic field.

An electric field diverges from a source.

A magnetic field curls around itself, and its divergence is zero.

There are no magnetic monopoles. A magnet will always have a North side followed by a South side.

An electric field can be monopoles such as a single negative particle.

When a positive particle moves in a region of magnetic field it experiences a force as shown in the right-hand rule:

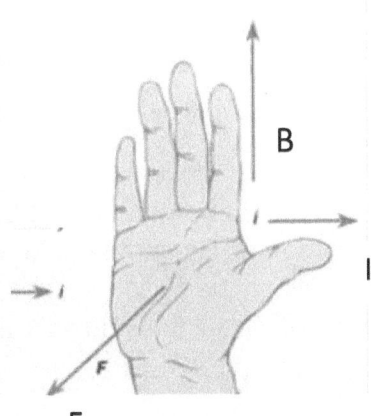

If the particle is negative the direction of the Force is opposite of what is shown:

Four fingers in the direction of Magnetic Field, Palm in the direction of Force, and Thumb in the direction of Current.

The two ways that the right-hand rule can be used is with the thumb pointing in the direction of the current I, the four fingers in the direction of the Magnetic Field, and the palm in the direction of the Force which applies for the flow of positive charges. For negative charges the Force will point in the opposite direction.

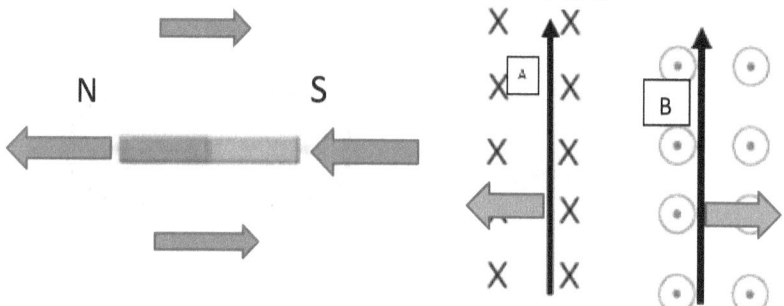

Black arrows in the direction of current in a wire

Green left and right arrows in the direction of Force

From a magnet the Magnetic Field points away from the North Pole and towards the South Pole.

Another use of the right hand rule is for coils where the four fingers point in the direction of the current, and the thumb in the direction of the Magnetic Field. A dot is used for a Magnetic Field out of the page while an x is used for a Magnetic Field into the page.

When a current runs through a coil it generates a magnetic field as shown:

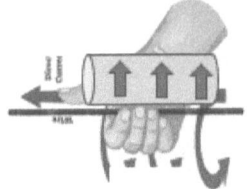

That is using the right-hand rule once again

Thumb in the direction of Magnetic Field and Four Fingers in the direction of Current in Coil.

Magnetic Field from a Solenoid.

An Electric Motor uses current through a wired loop in a region of a Magnetic Field to cause a Force on both sides of the loop with current in two different directions leading to a rotation. To allow the rotation to continue beyond 90 degrees the source of current reverses the direction of the current causing the rotation to continue as long as the Alternating Current continues to flow in the loop.

A wire with current in the direction I in a region with a Magnetic Field will experience a Force F. This is the basis of how Electric Motors work as shown.

That is how an electric motor works:

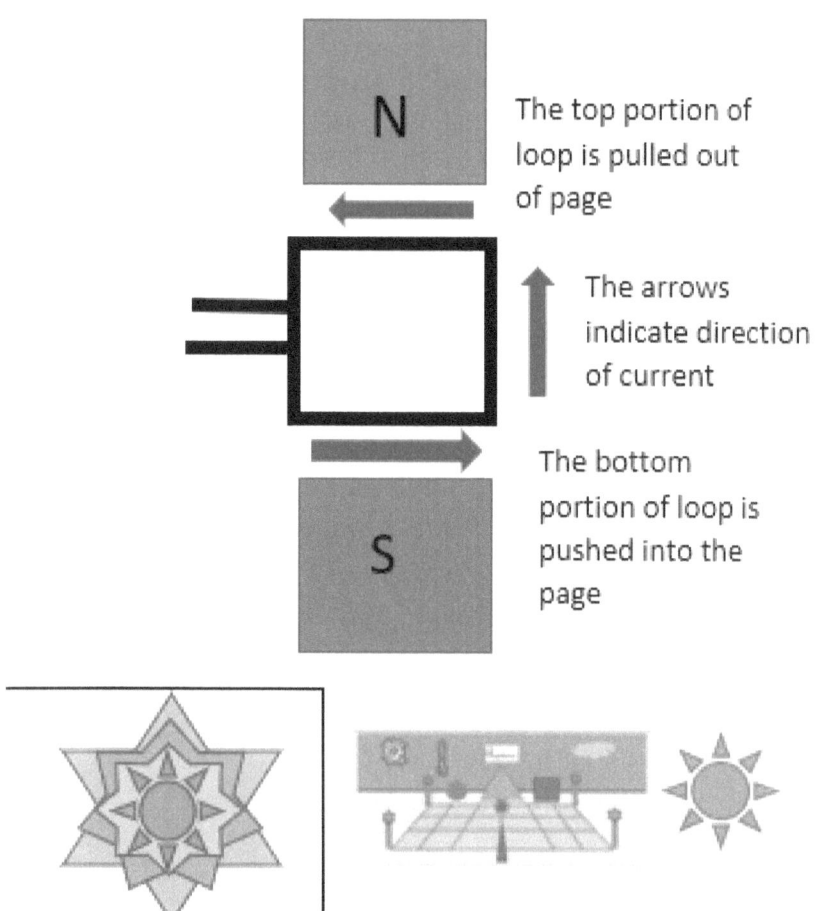

The top portion of loop is pulled out of page

The arrows indicate direction of current

The bottom portion of loop is pushed into the page

Induction Practice

Find the induced current in the following examples:

1...Coil flat on a region of magnetic field out of the page. The field decreases: <u>**CURRENT COUNTERCLOCKWISE**</u>

2.... Coil flat on a region of magnetic field out of the page. The field increases: <u>**CURRENT CLOCKWISE**</u>

3… Coil flat on a region of magnetic field into the page. The field decreases: **CURRENT CLOCKWISE**

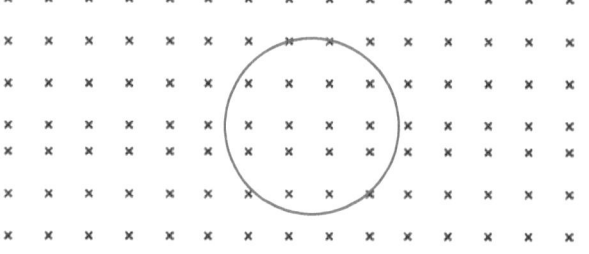

4… Coil flat on a region of magnetic field into the page. The field increases: **CURRENT COUNTERCLOCKWISE**

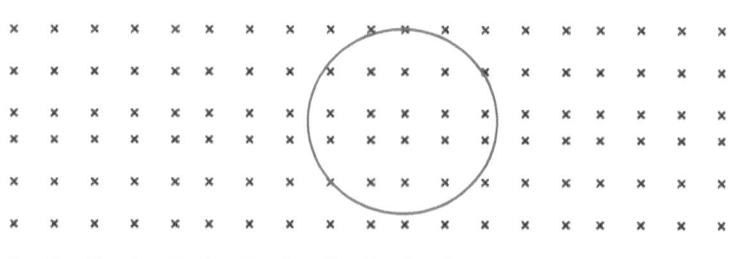

5…Coil travelling from the left into a magnetic field out of the page: **CURRENT CLOCKWISE**

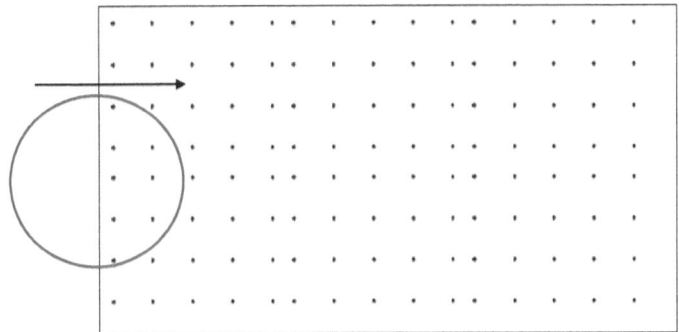

6…Coil travelling to the right leaving the magnetic field out of the page: **CURRENT COUNTERCLOCKWISE**

7… Coil travelling from the left into a magnetic field into the page: **CURRENT COUNTERCLOCKWISE**

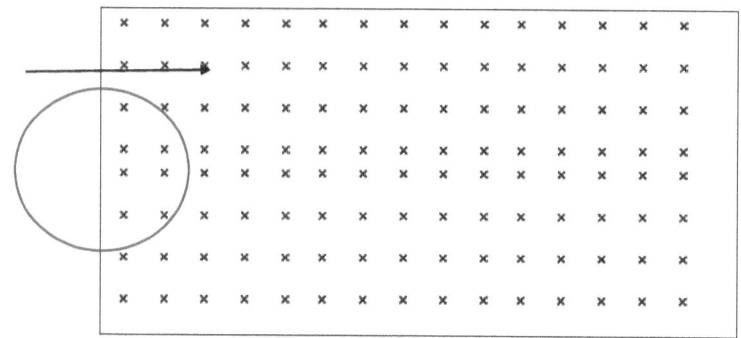

8… Coil travelling to the right leaving the magnetic field into the page: **CURRENT CLOCKWISE**

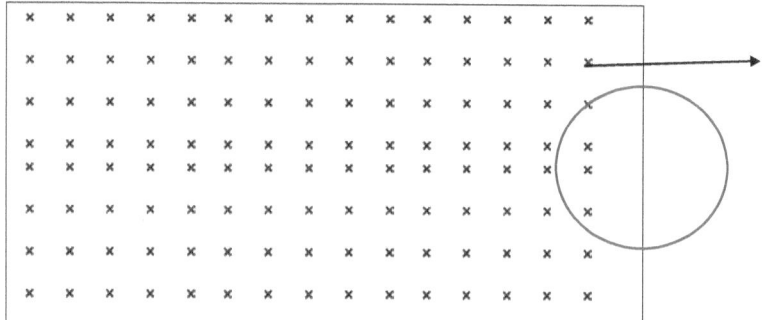

9…Coil travelling inside a region of magnetic field to the right: **NO CURRENT**

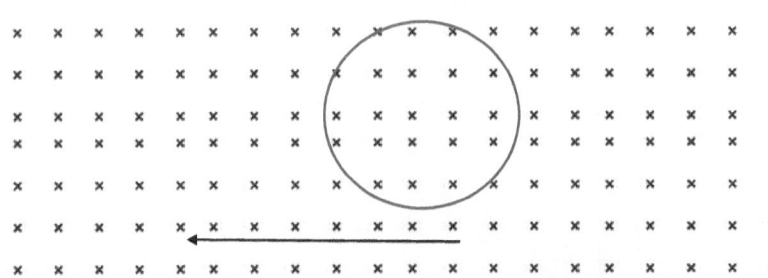

10… Coil travelling inside a region of magnetic field to the left: **NO CURRENT**

11… Coil travelling inside a region of magnetic field to the right: **NO CURRENT**

12… Coil travelling inside a region of magnetic field to the left: **NO CURRENT**

What is the induced current on the other coil?

A…

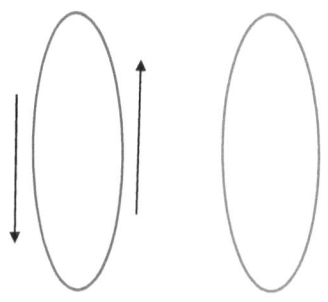

With a pulse of current in one loop there is a generation of an opposite current induced in the other loop.

B…

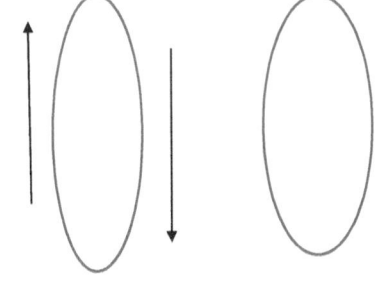

Continuous induction can only occur through pulses or changes in the current in one loop that affects the other at that given instant.

C…

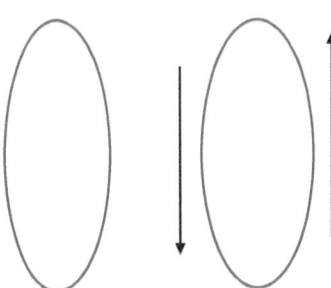

D...

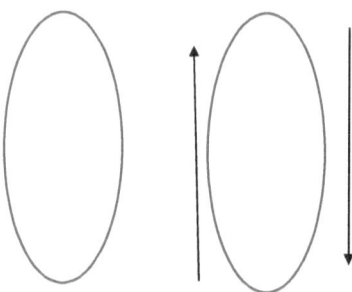

If the pulse of current in one coil is clockwise the induced current on the other loop will be counterclockwise and if the pulse is counterclockwise the induced current on the other loop is clockwise.

Vloudel then opened another book and read about forces due to current in a Magnetic Field.

When a wire carries a current a Magnetic Field is generated that curls around the wire like in the figure below using the right hand rule.

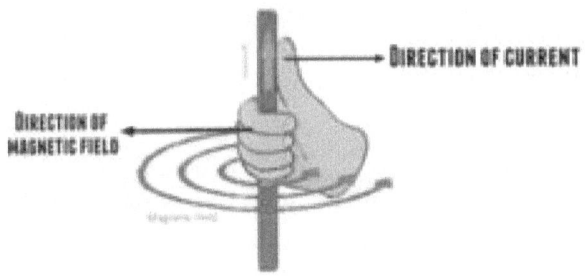

This phenomenon is explained by the Maxwell Equation $\nabla \times B$ which states that a change in Electric Field generates a Magnetic Field. When a current is flowing through the wire a Magnetic Field curls around the wire.

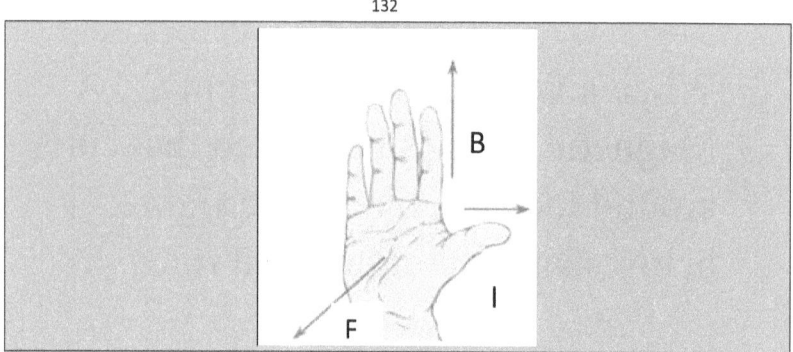

When a current flows through a wire that is present in a region of Magnetic Field, the direction of the force can be calculated by using Fleming's Left Hand Rule or the Right Hand Rule as show above.

Vloudel then showed Zeno some of the worksheets in Ogo's Book with answers:

| Label the direction of the forces on the wires with current flowing in the direction of the arrows exposed to the magnetic field as shown: | An x shows Magnetic Field into the page and a dot shows a Magnetic Field out of the page. |

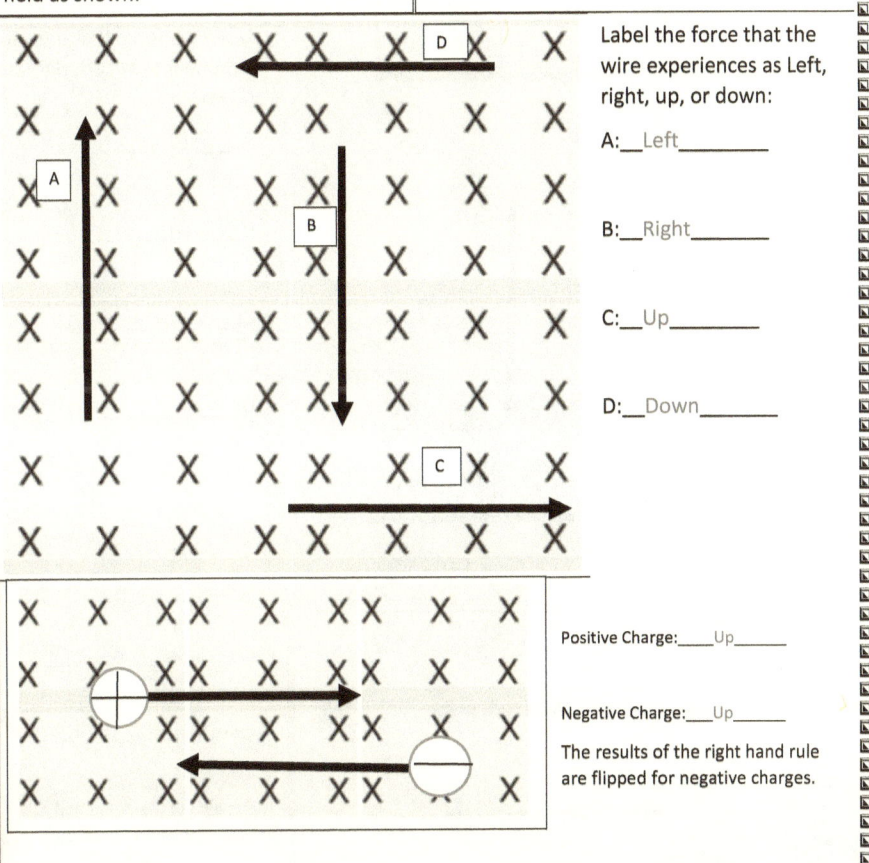

Label the force that the wire experiences as Left, right, up, or down:

A:__Left_____

B:__Right_____

C:__Up_____

D:__Down_____

Positive Charge:____Up_____

Negative Charge:____Up_____

The results of the right hand rule are flipped for negative charges.

134

Label the direction of the forces on the wires with current flowing in the direction of the arrows exposed to the magnetic field as shown:

An x shows Magnetic Field into the page and a dot shows a Magnetic Field out of the page

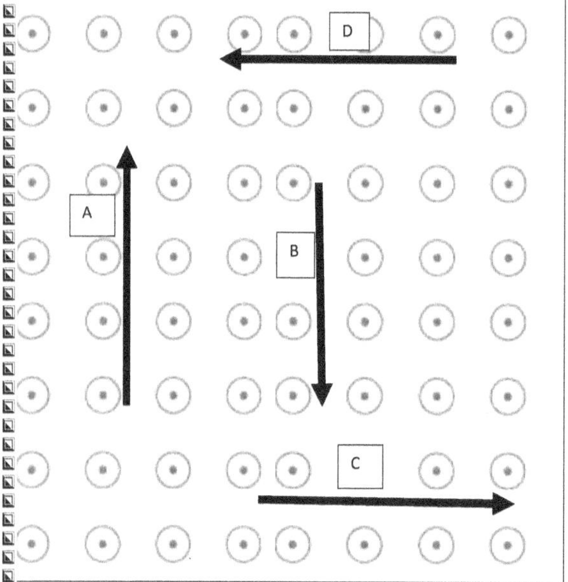

Label the force that the wire experiences as Left, right, up, or down:

A:___Right_____

B:___Left_____

C:___Down_____

D:___Up_____

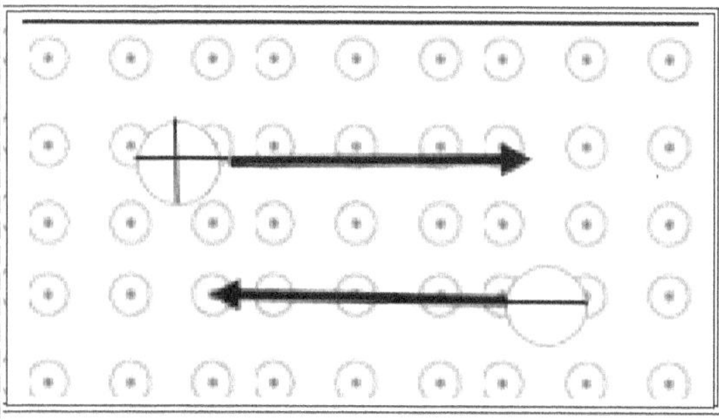

Positive Charge:___Down_____

Negative Charge:___Down_____

The results of the right hand rule are flipped for negative charges.

Next Vloudel showed Zeno some pages on optics:

Convex and Concave Mirrors and Lenses

Optics

Light is an electromagnetic wave. It is composed of an electric and a magnetic field that oscillates at 90 degrees from each other.

The electric field $= \frac{Magnetic\ field}{\sqrt{\mu_0 e_0}}$

Velocity of light $= \frac{1}{\sqrt{\mu_0 e_0}}$

Mirrors

Concave mirrors:

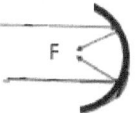

Converges light in front of mirror. Focal length is positive and real. The image can be positive real or negative virtual.

Convex mirrors:

Converges light behind mirror. Focal length is negative virtual. Image is always negative virtual.

Reflection of Light:

Light reflects off surfaces at the same angle as the incident angle. The angle is measured from the normal.

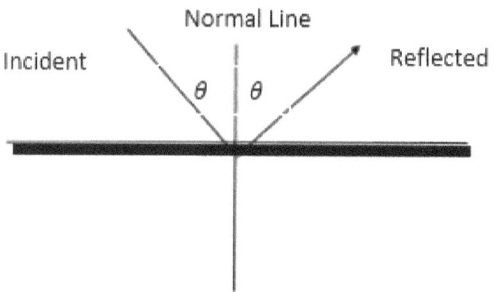

The equation for mirrors and lenses is:

$$\frac{1}{object's\ distance} + \frac{1}{image's\ distance} = \frac{1}{focal\ length}$$

The image and focal length can be negative or positive depending if they are virtual or real.

The focal length is where light from infinity converges.

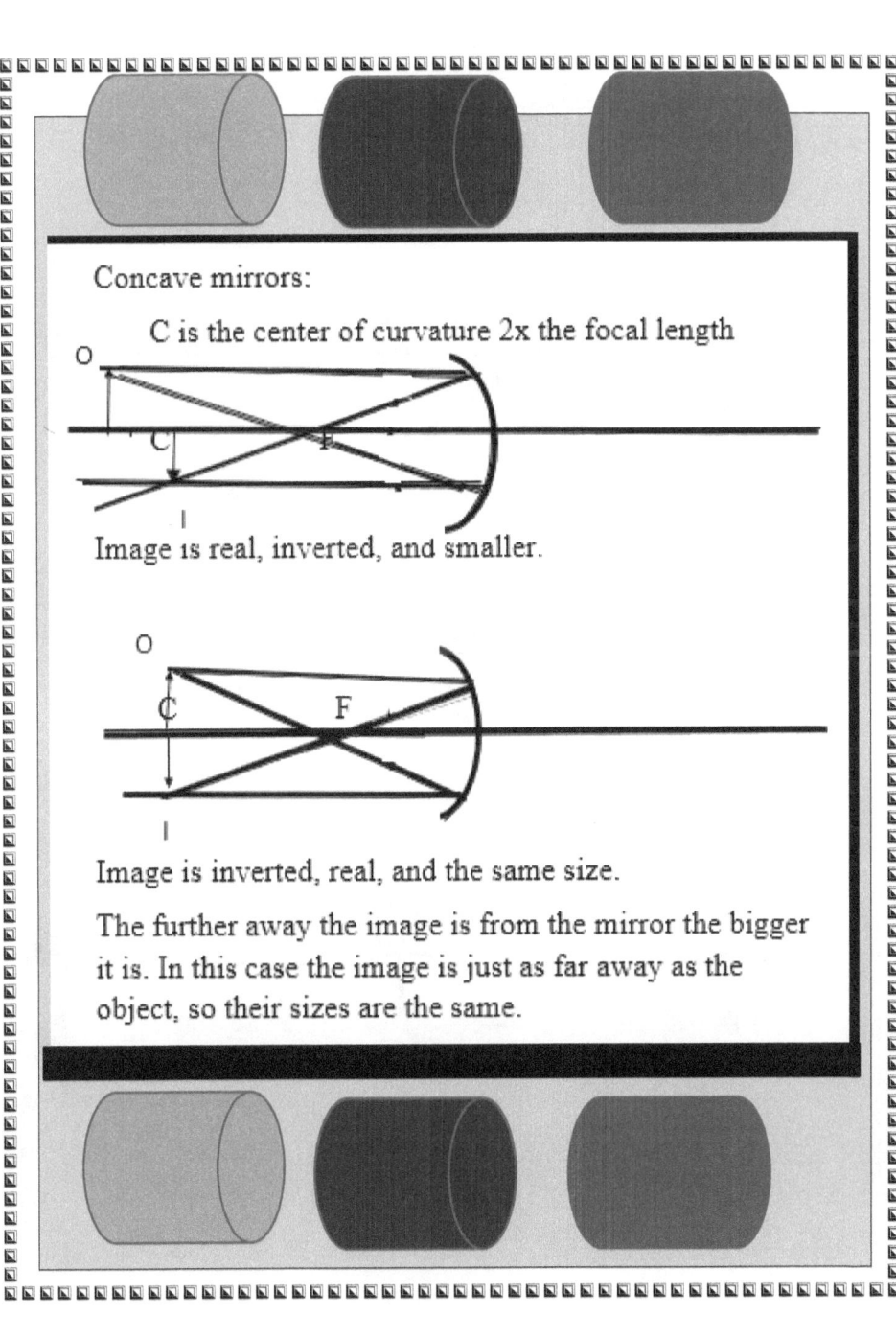

Concave mirrors:

C is the center of curvature 2x the focal length

Image is real, inverted, and smaller.

Image is inverted, real, and the same size.

The further away the image is from the mirror the bigger it is. In this case the image is just as far away as the object, so their sizes are the same.

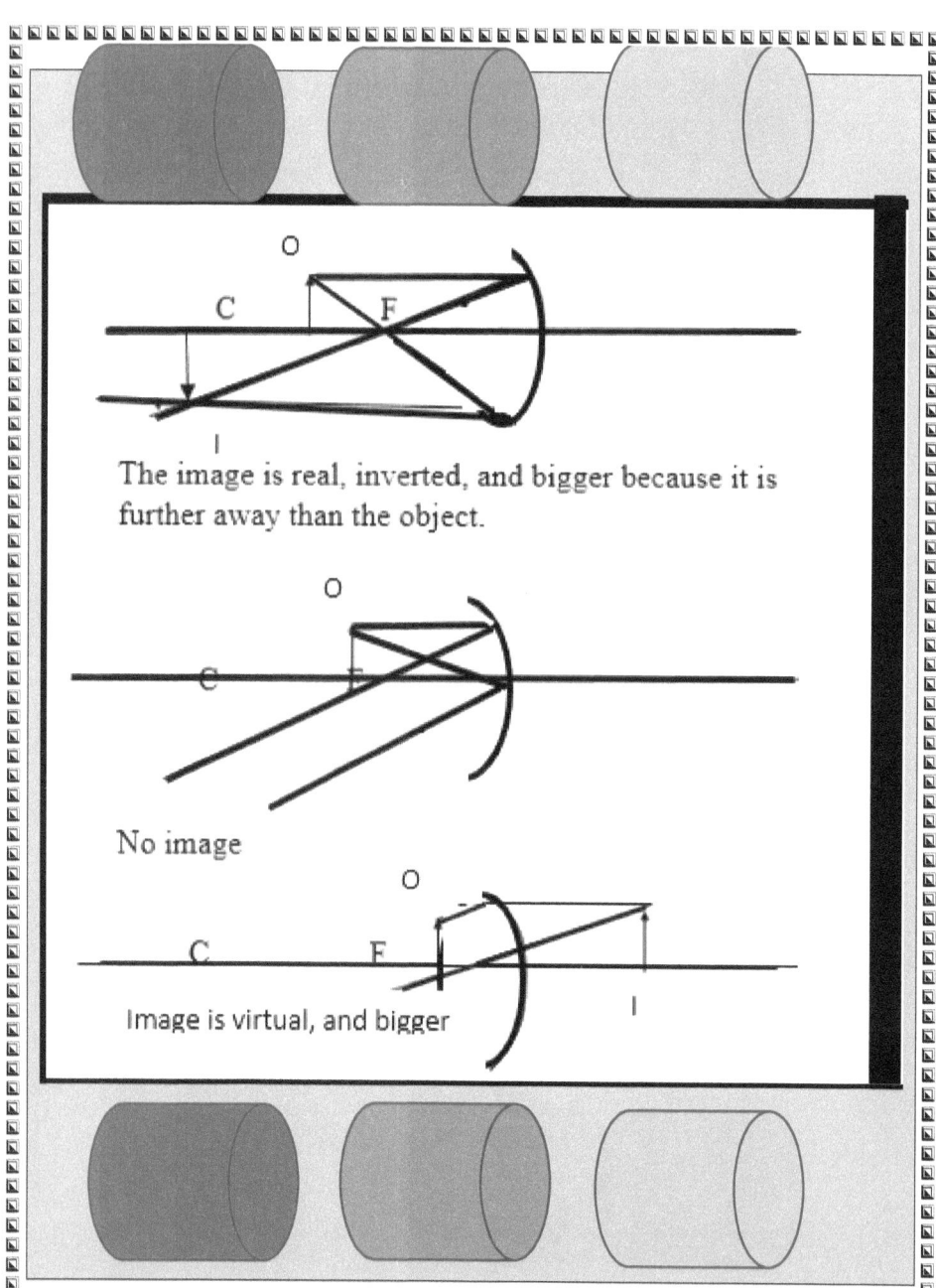

Convex mirrors:

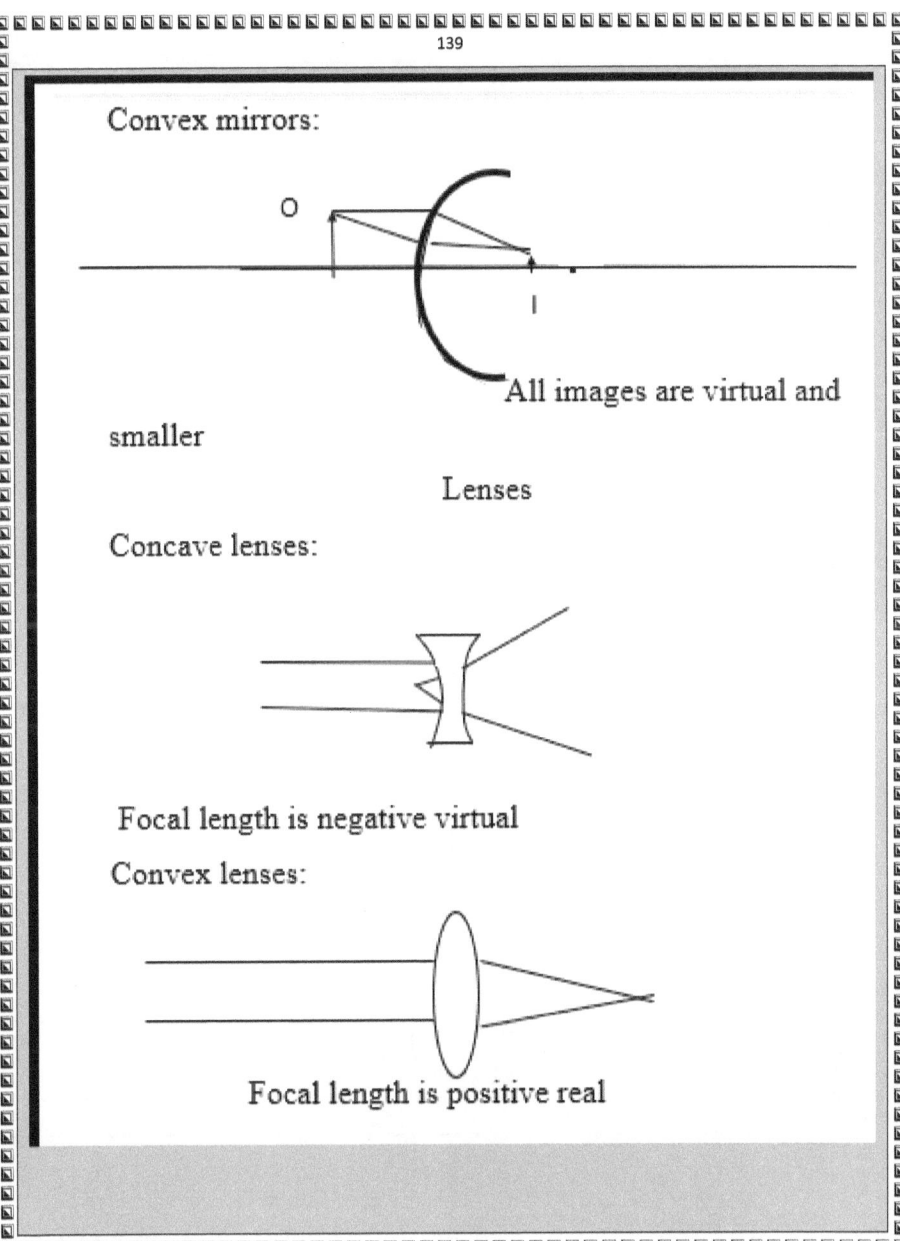

All images are virtual and smaller

Lenses

Concave lenses:

Focal length is negative virtual

Convex lenses:

Focal length is positive real

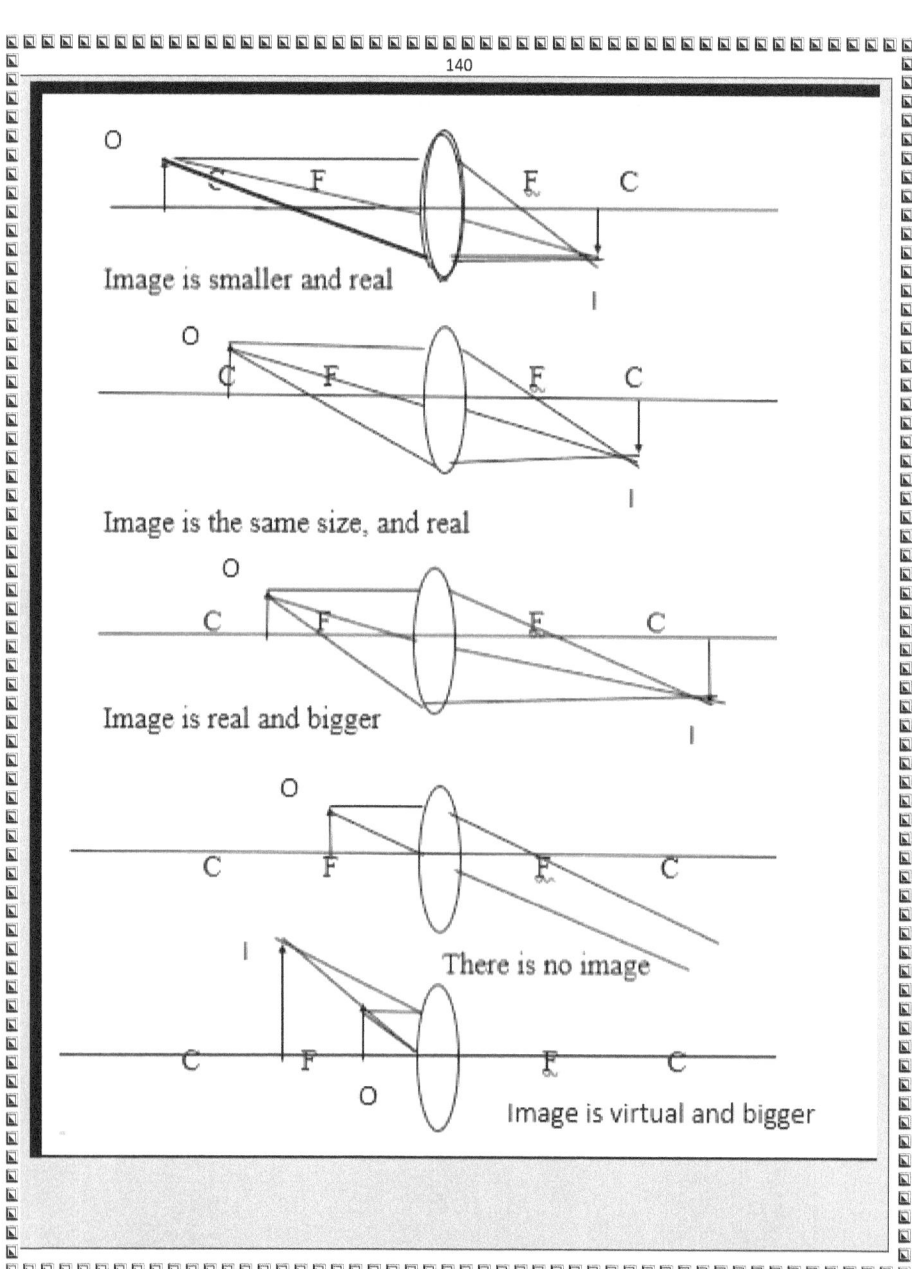

Image is virtual and bigger

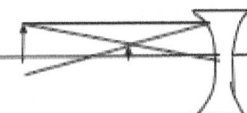

For concave lenses the image is always virtual and smaller.

Magnification of images $= \frac{-image\ location}{object\ location} = \frac{-Height\ of\ image}{Height\ of\ object}$

Magnification is positive if the image is virtual.

If the image is virtual it is right side up.

If the image is real, it is upside down, and its magnification is negative.

When light goes through two slits it causes an interference on a screen after the two slits.

Constructive interference causes the visible light patches that are followed by destructive interference that are the dark patches.

Constructive interference occurs at the following angles:

$Sin\theta = \frac{m\lambda}{D}$ where m = 0,1,2,3.....

Where D is the distance between the two slits and λ is the wavelength of the light.

Destructive interference occurs at the following angles:

$Sin\theta = (m + 0.5)\frac{\lambda}{D}$, where m = 0,1,2,3....

When light goes through a single slit, it also produces an interference. The location of its dark bars is at the following angles:

$Sin\theta = \frac{m\lambda}{W}$ where m=1,2,3…and w is the width of the slit.

When using a telescope, the minimum angle where it is possible to distinguish 2 points separately is:

$Sin\theta = \frac{1.22\lambda}{D}$,

Where D is the diameter of the telescope's aperture, and λ is the wavelength of the incoming light.

Waves are Everywhere

Explanation for Double Slit Experiment

Double Slit Interference Explanation:

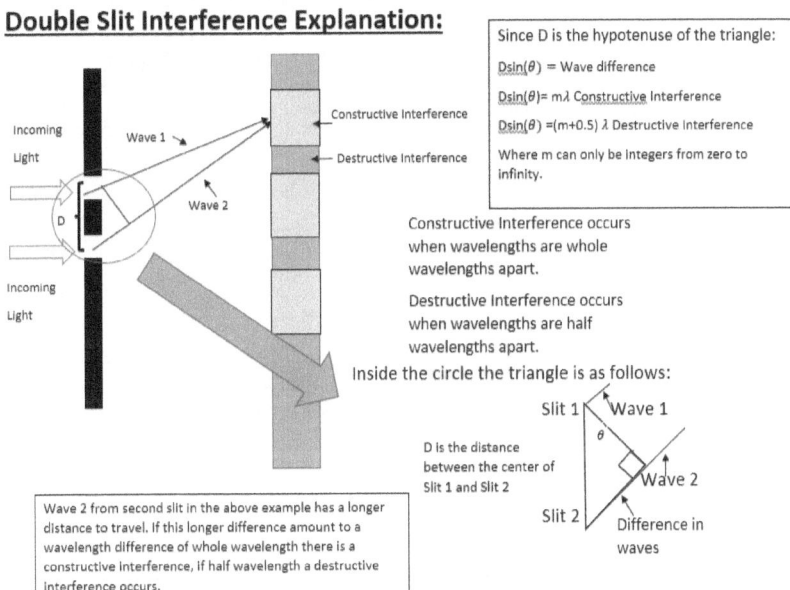

Since D is the hypotenuse of the triangle:

$D\sin(\theta)$ = Wave difference

$D\sin(\theta) = m\lambda$ Constructive Interference

$D\sin(\theta) = (m+0.5)\lambda$ Destructive Interference

Where m can only be integers from zero to infinity.

Constructive Interference occurs when wavelengths are whole wavelengths apart.

Destructive Interference occurs when wavelengths are half wavelengths apart.

Inside the circle the triangle is as follows:

D is the distance between the center of Slit 1 and Slit 2

Wave 2 from second slit in the above example has a longer distance to travel. If this longer difference amount to a wavelength difference of whole wavelength there is a constructive interference, if half wavelength a destructive interference occurs.

When the difference in wavelength is a whole number of wavelengths there is a constructive interference since the waves add up. When the difference in wavelength is a half wavelength the waves cancel and there is a destructive interference.

Vloudel then gives Zeno a tesla coil from a mysterious bag. Vloudel also picks up a light bulb from the bag and moves it closer to the tesla coil. Zeno then watches the light bulb turn on without wires.

Vloudel: Do you see this Zeno? It is possible to light up a bulb wirelessly. This means that there must be a field of particles in this region of space coming from the Tesla Coil that near the bulb gives it light.

Zeno: Amazing! Explain to me fields!

Vloudel: Electric Fields for example is a sea of particles in space that can generate a force. When a charge is placed in a field it experiences a force from the field.

Zeno: Are we under the influence of a field right now?

Vloudel: Yes. There are particles all over the universe and there simply is no escape from it. They are everywhere.

Zeno: That is so cool!

Vloudel: Yes Zeno but now it seems like we are done with today's lesson unfortunately.

Zeno: At first I was no liking it, but now it is time to leave?

Vloudel: Yes before you get bored again. It is good to leave when things are still going well.

Vloudel then pulled a rod from the bag and drew the entrance of the gray portal in space which then became solid and real. The two that floated were then pulled through the portal and Zeno returned to the Sahara Desert but did not see Vloudel anymore. He looked all around him and could not find that strange creature. He was then once again alone with his camel and the sun was setting under the dunes as the wind blew. He seemed to have been out the entire day and feeling very tired he sat on the ground and looking at the show of lights at the end of the day he said: **Good bye sun….**

And fell asleep….

Starcast Arcturus

Zeno

© 2021 Diogo de Souza
All Rights Reserved.

Contact Information:
diogodesouza7@gmail.com
diogodesouza7@hotmail.com

Zeno' adventure in the Middle East:

Zeno walked across the desert in Southern Iraq with his backpack filled with books that he acquired from his trips around the world when he arrives at the great ziggurat from the ancient city of Uruk. He was marveled seeing this great structure at a distance and walked faster anxious to arrive at the location. Upon arrival a desert nomad stopped him and asked: "Where are you coming from?

Zeno: I come from foreign lands and I travel throughout the world seeking the knowledge of all things, and I carry a backpack with me containing some parts of everything that I know."

Nomad: Haven't you heard the news lately?

Zeno: What do you mean?

Immediately a great light showed up in the sky like a burning torch coming rapidly through the atmosphere and everyone in the desert seeing the great show of light became desperate and decided to run for their lives.

Nomad: As you can see we got to run.

They did not have enough time and the great object hit the ground at a distance and exploded with a great mushroom of smoke from the ground up to the air.

Zeno: We need to go there to see what that thing is.

Nomad: Bad idea. What if it is radioactive?

Zeno pulled out of his bag a small radiation detector and said:

Zeno: Then we can go back and ask someone else to do it for us.

The two walked towards the smoke that was slowly spreading in the air and beginning to fade. They walked toward the smoke while several people ran in the other direction, away, and yelling by telling the two to quit and run away as well.

Zeno: It is all under control. My radiation level still reads low levels of radiation just slightly higher than normal.

Nomad: I hope that this idea of yours is worth it. I am sorry but I am leaving. Good luck for you.

The nomad left Zeno and Zeno walked among the smoke by himself following a GPS on his smart watch towards the location of the collision.

Zeno then arrived in the scene and met the object. It looked like a spaceship in the shape of a disc. The door was open and Zeno walked its steps towards the inside.

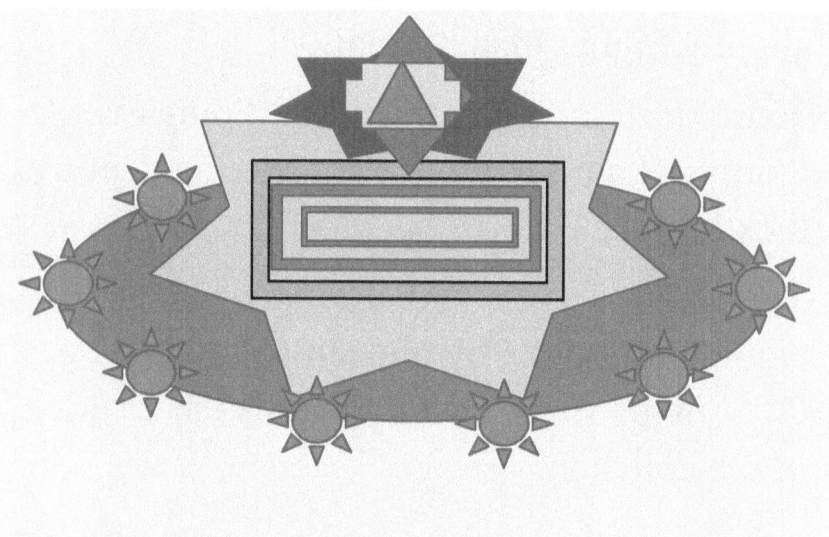

While inside Zeno saw several instruments on a panel and blinking lights and on the pilot's seat a creature asleep possibly due to concussion. Zeno realized that it was his old friend Uflonix.

Zeno: Uflonix wake up!

Zeno pulled Uflonix and by slightly hitting his face with his hands, he was able to bring Uflonix back.

Uflonix: What is happening here? Where am I?

Zeno: You are near the ancient city of Uruk. I saw your ship falling uncontrolled across the sky towards the ground.

Uflonix: Oh no! I was actually heading to Europe but anyways. Please grab me that book over there.

Uflonix pointed at a book behind him and Zeno went to pick it up.

Zeno: What about the book?

Uflonix then grabbed a rod and he hit on the head of Zeno who was holding the book and immediately Zeno disappeared.

When Zeno opened his eyes extremely surprised he was in what it looked like a planetarium. He was also invisible. He could not see any of his body parts, not even his arms. He could not see himself but he knew that he was there because he could see everything around him and feel it. He seemed like a ghost in empty space and he contemplated the Milky Way across the sky.

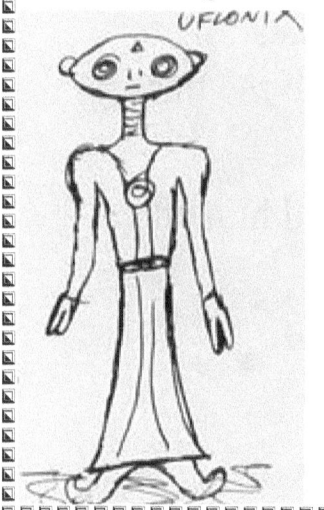

Uflonix then began to speak with Zeno. Zeno could hear Uflonix's voice but also could not see Uflonix:

Uflonix: Let me introduce you to Astronomy my favorite topic after so many of my trips across the Milky Way.

SERIOUS
ASTRONOMER

Diogo Franklin de Souza

Introduction:
What is Astronomy:

Astronomy is the study of the cosmos. Since everything we see around us is part of the universe, somehow Astronomy is the study of the entire natural structure where we live in. This structure is the universe, a system that began a long time ago through mysterious causes, and with a future that is very uncertain. Maybe you will help science in some new discovery.

Scientists today agree with the fact that the universe is some 13.5 Billion of Earth years old, whose cause of origin is an unsolved mystery. We think that the universe was very small in the beginning, like smaller than a Proton, or at least that is what is believed. The universe then had a rapid expansion from which Atoms, Stars, and Galaxies formed as this cosmic bubble became cooler. The bigger the inflating universe, the cooler and less dense it became. This led to the formation of all the matter that we see today.

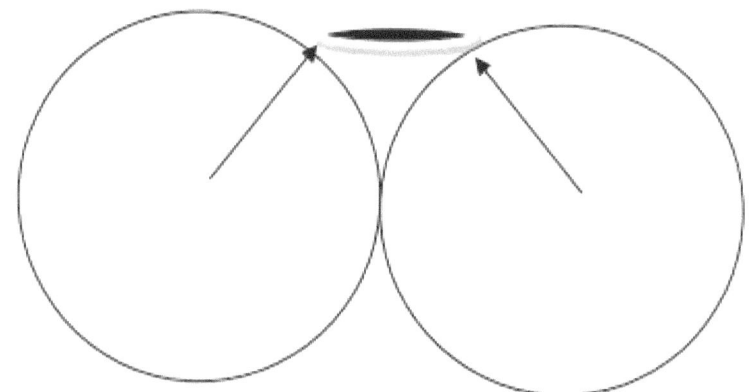

Stars have a life cycle. Several chemical reactions happen inside a star and the energy produced from these reactions gives a star its light. After combining a multitude of atoms and forming heavier ones, a star reaches the end of its life with an explosion called Supernova. After the explosion, the star may become a Black Hole if it has a large mass, a Neutron Star if it is less massive, or a White Dwarf if it has very little mass. When stars die, they explode giving out heavy atoms in space in clouds called Nebulas. These later come together forming other stars, producing star clusters, which later expand, forming stars and solar systems such as the one we live in. The Sun is not the first star, neither will be the last star in the universe. Thanks to other generation of stars in the past, the Sun and the Earth had enough heavy raw materials to allow life to form in this planet. We are star dust.

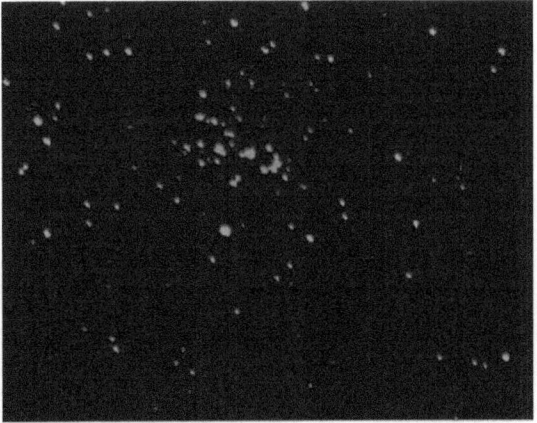

Our Sun is one star among billions in this Galaxy, and our Galaxy is one among billions in the universe. Life most likely can not just be found here but must exist in many other planets. We have not found life elsewhere in the cosmos, but common sense gives us a desire to explore as much as we can hoping that our reasoning is right.

Our Sun is one star in our Galaxy called the Milky Way, and our Planet Earth is one of the eight planets in our Solar System. The sky we see in a clear night, tells much about our history, and about the cosmos. We need to explore this wonder, and that is the purpose of science towards the progress of our understanding.

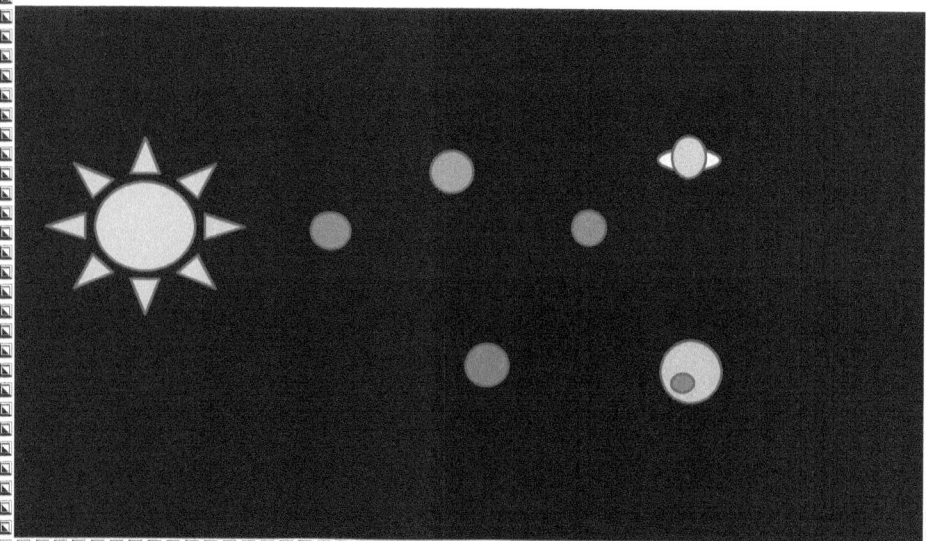

Observing the Sky:

You go outside in a clear night with a compass. You face the northern direction, and then south, east, and west. You should notice that the stars, the Moon, the Sun, the Planets rise from the east, and set on the west. So now you know that the sky is like a clock. Given the fact that the Earth rotates around its axis, the points in the sky seem to move eternally from east to west making a complete circle around the celestial globe in one day. The Earth is also revolving around the Sun, which causes stars to rise earlier in the east and set earlier in the west in increments of time per day. The Moon is revolving around the Earth from west to east in the same direction as Earth's rotation. This causes the location of the Moon to be more eastern each day. One complete revolution of the Moon around the Earth is about a month. As the Earth revolves around the Sun, the location of the Sun in the sky also changes. The Sun, the Moon, and the other Planets in the Solar System move in a line in the sky called the Ecliptic. They all seem to move from west to east, so after each day they rise in the east earlier than the day before except the Moon which appears later each consecutive day, and the Sun which just moves in front of the background of Stars and Constellation. Planets also seem to move from east to west for a while before continuing to advance towards the east. This happens because the Earth may pass the other Planet in its orbit around the Sun.

The point in the sky directly above a location on the Earth is called the Zenith. The complete circle in degrees surrounding a location on Earth is the Azimuth. Each location on the surface of the Earth will have a different Zenith, and a different location of the stars in the sky at an instant in time.

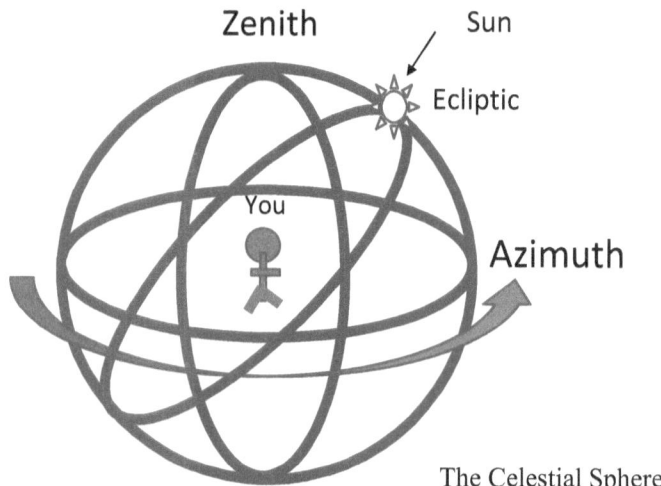

The Celestial Sphere

The 12 Constellations of the Zodiac:

The ecliptic is an imaginary line that runs through the 12 Constellations of the Zodiac. The Twelve Constellations are:

	Constellation	Brightest Star
1	Pisces	Circle of Stars under Pegasus
2	Aries	Hamal
3	Taurus	Aldebaran
4	Gemini	Pollux and Castor
5	Cancer	Stars forming an X shape with a cluster in the middle
6	Leo	Regulus
7	Virgo	Spica
8	Libra	A sideways triangle
9	Scorpius	Antares
10	Sagittarius	Near the center of the Milky Way Galaxy
11	Capricornus	A parallelogram on its Eastern Side
12	Aquarius	An X on the Northern Part

The Sun, Moon, and Planets can only be found along the Ecliptic.

The seasons of the Earth are determined by the location of the Sun along the Ecliptic. When the Sun's location on the Ecliptic is north of the Equator, it is warmer in the Northern Hemisphere. When the Sun is South of the Equator, it is warmer in the Southern Hemisphere. The Sun is not what is moving, but it is just its location in relation to Earth.

The Zodiac

The Earth's tilt causes the location of the Sun in the sky to change through seasons. There are four seasons:

Northern Hemisphere

Season	Begins
Summer	June 20, 21
Fall	September 22, 23
Winter	December 21, 22
Spring	March 20

Southern Hemisphere:

Season	Begins
Summer	December 21, 22
Fall	March 20
Winter	June 20, 21
Spring	September 22, 23

The Ages

The Sun's location in the Ecliptic along the Zodiac on March 20 defines the age we are in. Currently we are in Pisces and heading towards Aquarius. That is because the Earth suffers precession and the location of the Sun at a given time in a year will not be exactly the same as the year before. Human civilization began in the age of Taurus, and then came the age of Aries, and Pisces where we currently are. We should enter the Age of Aquarius in a few centuries after 2020. The Sun's location in the sky on March 20 is called the Vernal Equinox for the Northern Hemisphere.

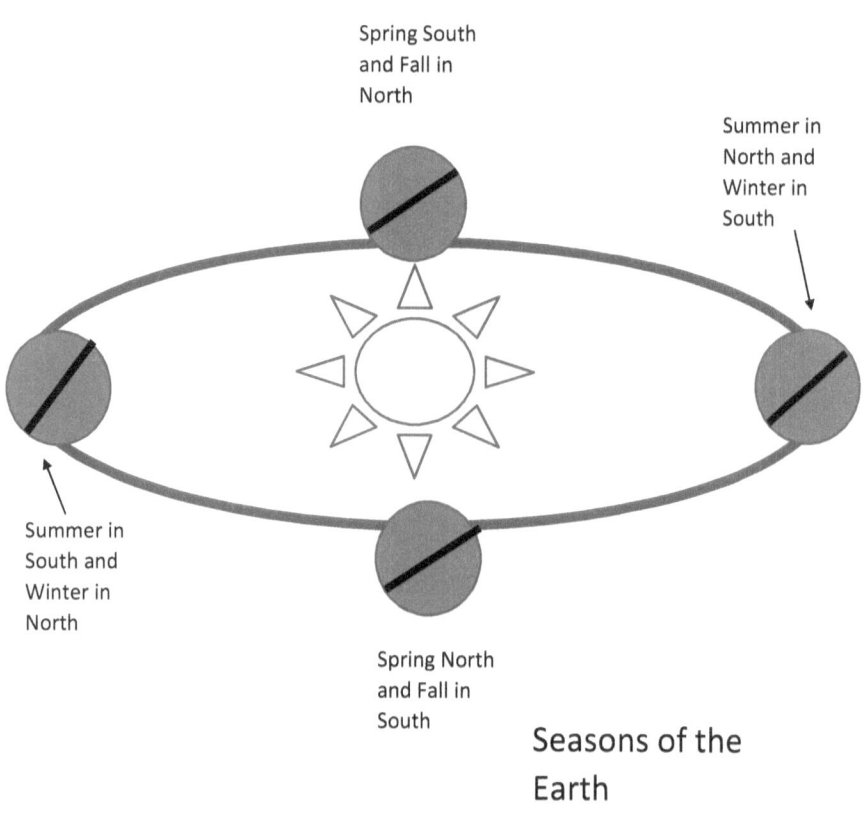

Seasons of the Earth

The Phases of the Moon:

As the Moon revolves around the Earth it undergoes phases. These phases are helpful in determining the length of the month. There are 12 months in a year.

A Solar Eclipse happens when the Moon covers the light of the Sun as seen from the Earth. A Lunar Eclipse occurs when the Earth covers the Sunlight on the Moon as seen from the Earth.

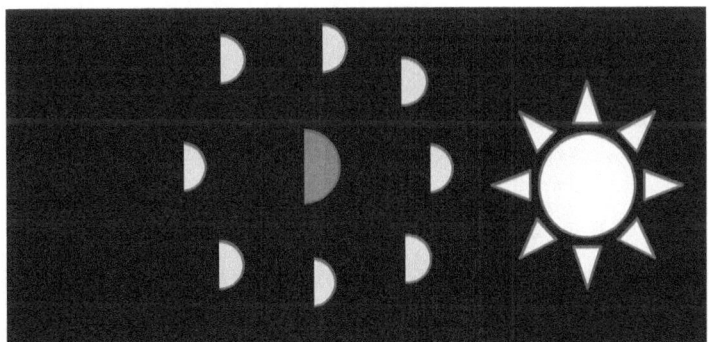

Phases of the Moon

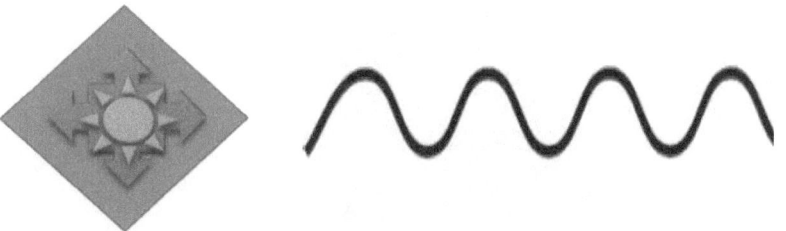

The Telescope:

There are three kinds of telescopes:
Refractor Telescopes that use lenses in all of its optics.

Reflector Telescopes that use primary and secondary mirrors before the light arrives at the eyepiece.

Cassegrain Telescopes that also uses mirrors but the light inside the telescope is reflected more times. This generates a longer Focal Length. It is a great Telescope for high magnification.

Equations

A telescope magnification is calculated by dividing the Telescope's Focal Length by the Focal Length of the eyepiece in use.

A telescope of Focal Length 1,000 mm can magnify 100 times with an eyepiece of 10 mm.

The maximum magnification of a telescope is found by multiplying the aperture in inches by 50.

A telescope that has a 4.5 inches diameter aperture can magnify up to 4.5(50) = 225 times.

The wider the telescope, the more light it can gather. A faint star for a narrow telescope can be a bright star for a wider telescope. Wider telescopes can see more stars and be able to see fainter deep sky objects such as Galaxies and Globular Clusters.

A Good Telescope:

To see the universe through a telescope it is preferred a very clear night in a location very far from light pollution. A small telescope with Focal Length of 400 mm and a 72 mm diameter aperture is ideal to observe wide views of an entire constellation and stars through a 40mm eyepiece. A telescope with a long Focal Length (longer than a 1,000 mm), and wide aperture, (wider than 90mm) is very useful to see Binary Stars, and Open Star Clusters, as well as deep space objects, such as Galaxies, Nebulas, and Globular Clusters.

What to look through a Telescope:

Without a fancy camera and much experience in Astronomy, you could start your observations by hunting for Binary Stars using the tables in this book. It is best to find these Binary Stars with a Telescope that offers a good magnification. You can also use a very small telescope to see entire Open Star Clusters such as the Pleiades, Hyades, and Perseus. The view is beautiful. You should use small telescopes for wide views, and bigger telescopes for greater magnification. Cassegrains are my favorite telescopes for high magnification while small FirstScope Reflector Telescopes are ideal for wide views of Constellations. That is why I have more than one telescope. I use them for different reasons.

Binary Stars:

Most stars in the universe are Binary. They are two stars that revolve around their common center of mass. There are still many questions, such as whether life would be possible in a planet orbiting such stars.

Planets are pieces of rock, or gas that condensed to form spheres of matter that revolve around stars. Without any telescope, it is possible to see Mercury, Venus, Mars, Jupiter, and Saturn. They look like stars without a telescope, but with a telescope it is possible to see them as spheres. Uranus and Neptune require a telescope to be seen. Uranus looks like a pale blue sphere in space, while Neptune is a dark blue dot through a backyard telescope.

Planets, Moons, and the Sun:

Planet	Feature
Mercury	Red sphere
Venus	Phases
Mars	Red sphere
Jupiter	Cloud bands on the atmosphere and 4 visible moons

Saturn	Yellow sphere with a yellow ring, and a couple of moons around it.
Earth's Moon	Phases of the moon and craters on the surface (must use lunar filter)
Sun	Yellow with black spots (must use solar filter)
Uranus	Faint white bluish little sphere.

Phases of Venus:

Since Venus is closer to the Sun than the Earth, it is possible to see this planet having phases. It is also noticeable changes in the size of the planet as it moves closer and away from the Earth.

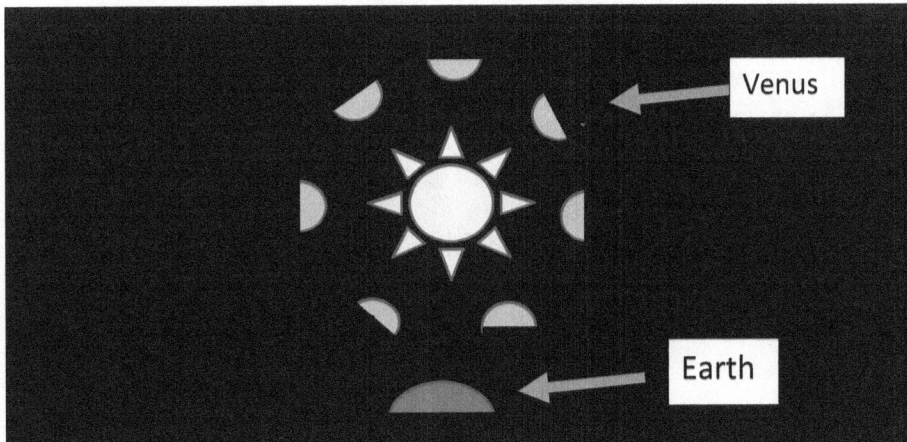

Venus appears almost fully lit when farther away and like a crescent when near.

Bright Stars:

Below is the table of the brightest stars seen from the Earth in the Northern Hemisphere.

Bright Star	Constellation
Arcturus	Bootes
Vega	Lyra
Sirius	Canis Major
Orion Belt (3 bright stars)	Orion
Rigel	Orion
Capella	Auriga
Deneb	Cygnus
Altair	Aquila
Procyon	Canis Major
Betelgeuse	Orion
Aldebaran	Taurus
Antares	Scorpius
Spica	Virgo

Another interesting star:
The North Star is **Polaris** which is right above the North Pole. It is located in Ursa Minor.

Interesting fact:

The center of the Milky Way is between the constellations **Sagittarius and Scorpius**.

The **Constellation Orion** is shaped like a guitar.

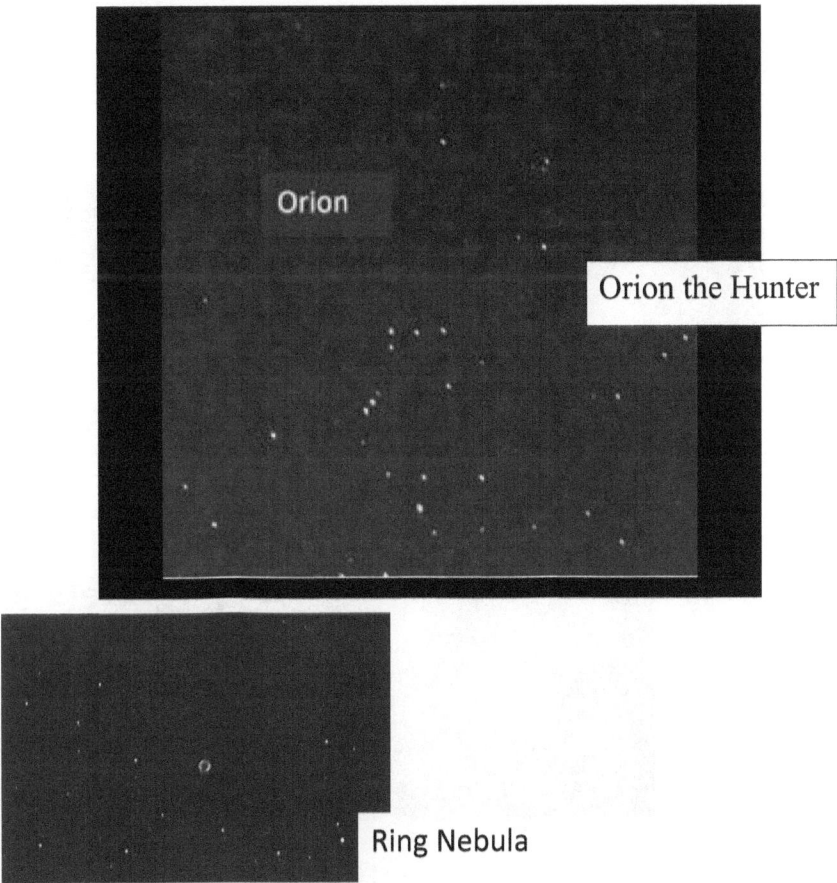

Some of the most beautiful targets in sky at night

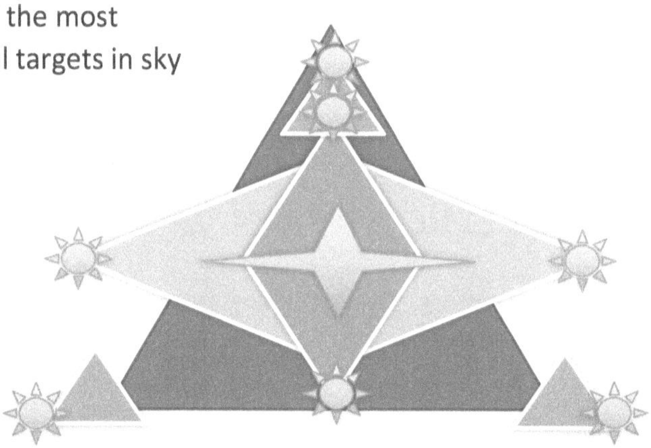

The Moon

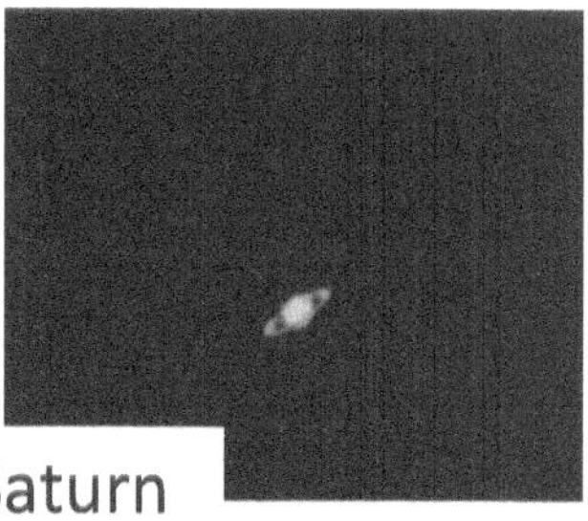

Saturn

Orion Nebula

The constellations with greatest number of visible stars in the Northern Hemisphere are:

1...Taurus
2...Perseus
3...Cygnus
4...Cassiopeia
5...Cepheus

Open Star Clusters for Binoculars:

1...Cepheus Star Cluster
2...Coma Berenices Cluster
3...Perseus Cluster
4...Pleiades
5...Hyades
6...Cancer Beehive Cluster
7...Orion Nebula
8...Cygnus Galactic Cloud
9...M7 Star Cluster
10...Orion 3 Stars Belt
11...Coat Hanger Star Cluster
12...Lyra Stars
13...Aquila Stars

Favorite places in the sky:

1 Perseus Double Cluster
2 Orion Nebula
3 M 35 in Gemini
4 Orion Belt
5 Pleiades
6 Omicron 1 Cygnus
7 Cor Caroli Double Star
8 Andromeda Galaxy
9 M 7 near the center of galaxy
10 Hercules Globular Cluster
11 Beehive Cluster
12 Albireo Double Star
13 Cepheus Cluster
14 Perseus Cluster
15 M 41 Little Beehive
16 Hyades Star Cluster
17 Cat Eye Galaxy
18 Coat Hanger Star Cluster

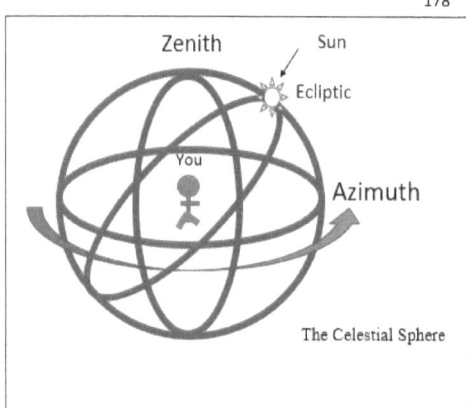

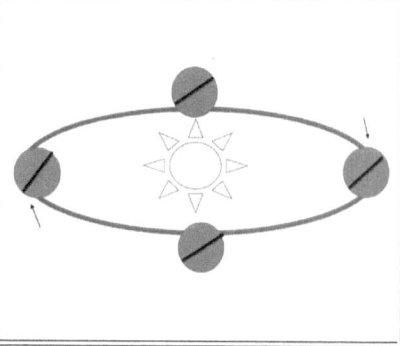

The celestial sphere, the explanation for seasons, and the star map for both south and north pole.

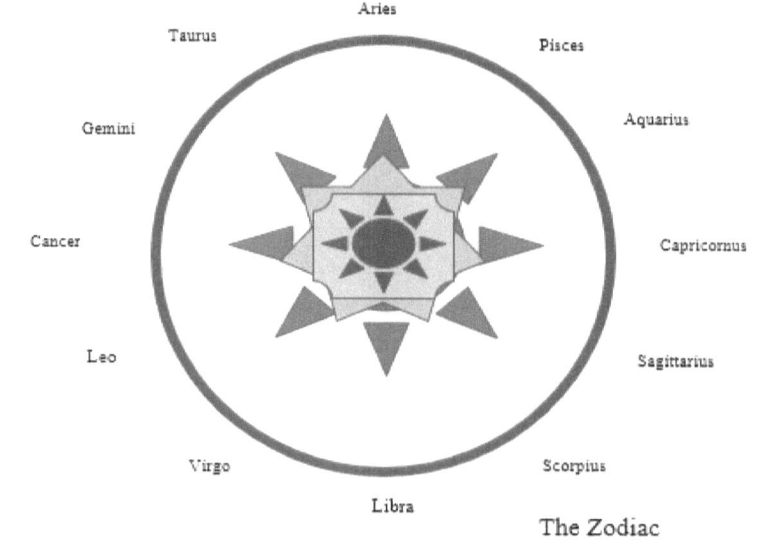

The Heavens are an immense clock slowly rotating

Uflonix finished showing Zeno his notes in Astronomy. They could not see each other and only hear each other. They could not even see their own bodies. They were invisible in the star filled space. Then all of the sudden they began accelerating towards a red star. The red star grew larger, and when closer they saw that it was not a star but rather the entrance to a tunnel. Uflonix said: "It is the Xynwheel!" After entering through the colorful portal their eyes were like spirals.

They were being transported to a parallel universe which was the universe of pure Mathematics. They were found in a sea of numbers, geometric figures, Mathematical Symbols all around in a crazy movement, integrals, summations, and they were marveled since it was an intense feeling.

Uflonix then spoke by reading from a scroll with his right index finger raised up:

Introduction to Philosophy

The larger the number of sides of a regular polygon, the greater is the definition of a circle.

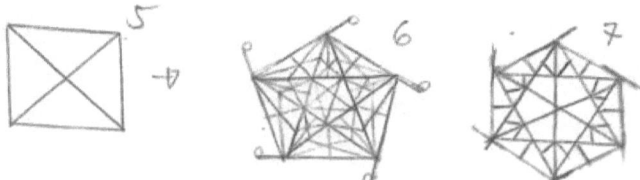

The quantum phenomenon gives structure to all that exist.

To study is to track the quantum.

To live science is to be always in search of something that is not yet known
or elaborate what is already known.

It is important to be aware of the quantum details of the universe, learning, teaching, and enjoying the time.

ART:

No scientific effort for the social and environmental benefit is in vain.

Art is a way to express ideas. It can change the world just like books do.

The libraries are a source of information and it has many books.

The books of a library are divided in themes.

Each theme and its books form the quantum sea of content about many things.

Time is made of small pieces that constitute seconds, minutes, hours, days, weeks, months, years, decades, centuries, and millenniums.

The numbers 0 to 9 are fundamental for all other numbers.

The quantum nature is made of small pieces that constitute the whole universe.

0 1 2 3 4

5 6 7 8 9

Pieces are everywhere. In maps there are illustrated representations that indicate parts of a system. In the universe there are laws and constants that make it possible for men to develop technology with the understanding of quantum pieces.

Science and Philosophy

Men have skills to perform deeds.
The human being needs to use those skills to do what is good.

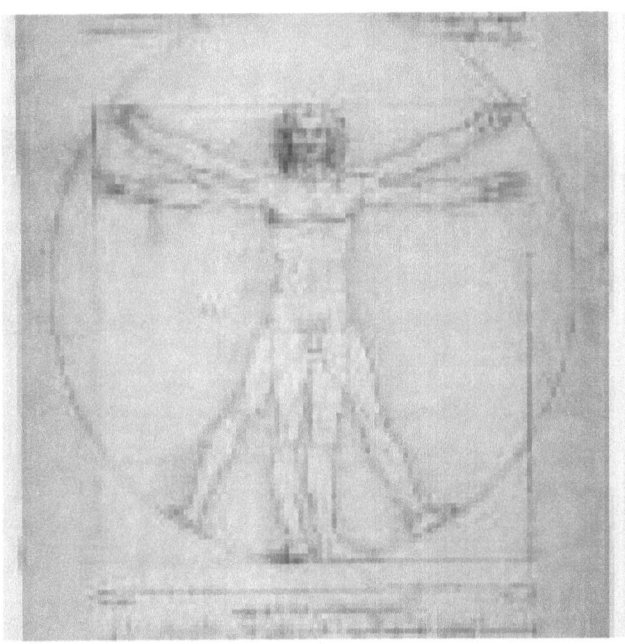

The pieces of the universe have limits that are within understanding.

When we study the universe, we study the limits established by the quantum laws and constants in the cosmos.

The universe can be measured and calculated because there is an order.

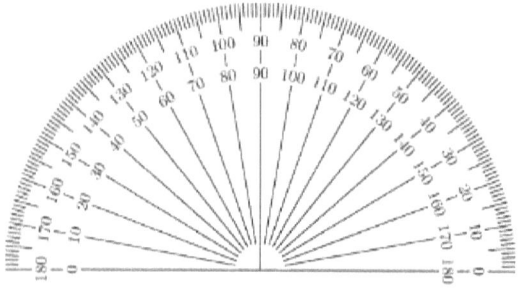

It is fun to study but everything needs a limit, and it is necessary to rest from deep study to learn better some other time.

There is a time for each thing.

There is a time to sleep, eat, study, and for physical exercise.

Time flows and many events happen.

The time for the universe has a beginning, middle, and an end.

The life of men and of all living beings has a beginning, middle, and an end.

In the universe there are limits, and these limits give structure to matter.

The material universe is provisory, and nothing in it is eternal. This is the law placed on everything.

The structure of the cosmos ages with time, becoming incapable of sustaining nature forever.

There is an end for everything in the material universe.

This limit is found in nature and it is irreversible.

Men can still have control over many aspects of nature.

There is a control of the cosmos in the electronic systems found in computers. These machines are built to be useful in communication through transfers of codified information. This and other forms of technologies prove that the limits of nature can be handled by men.

The universe is mechanical like a computer.

There is control over the cosmos in all technologies.

Understanding the cosmos makes it possible to have some control over matter in the development of technology.

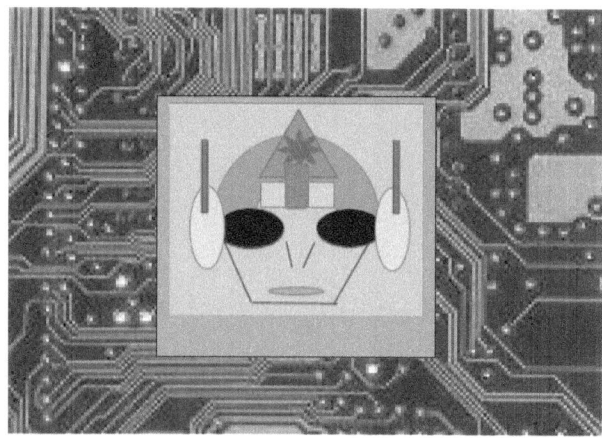

Inside the universal limits, much is possible technologically.

With a hammer in hand we have control of its position and velocity and consequently some control of the position and velocity of the particles of the composition of the hammer.

When we heat food in the microwave, we order the technology to increase the kinetic energy of the particles in the food.

It is possible with the aid of technology to control the temperature of the food.

With a refrigerator we keep food cold. With a microwave or a stove we heat up food. When making food we have control of its chemical composition and temperature.

In an electric circuit it is possible to direct moving electrons through the wire.

That way a lamp is illuminated, and in this system we have control over the cosmos. The lamp can be on or off as we wish.

When chemists make medications for patients, they have control of nature by controlling the chemical composition of the medicine.

A simple throw of a rock, a jump, or a walk, involves control of the cosmos.

All living beings have some control when moving themselves, moving objects, eating, and other things.

Of all living beings, men are the wisest.

Men have to know how to correctly control what it can for the benefit of the universe.

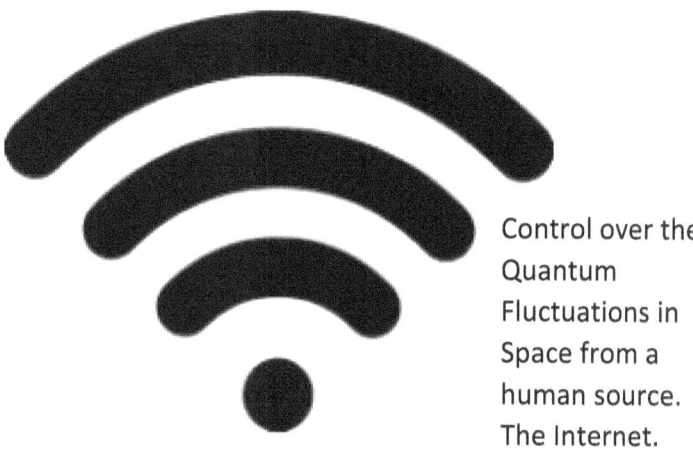

Control over the Quantum Fluctuations in Space from a human source. The Internet.

We have intelligence to discover all the mysteries of the cosmos.

There is one absolute truth, not many, that describes the universe. The truth is simple, easy to understand. It is knowledge to see something hard in a simpler manner to facilitate comprehension.

All chemical reactions that we cause prove our power to know the universal proprieties. We can do many things inside our limits.

Mathematical principles are laws and constants used to define and understand cosmic aspects.

Mathematics shows that there exists an order for everything.

It is possible through mathematics to understand all natural phenomena.

Having an order in the chaos it is somewhat possible to predict the weather based on the present, and the natural order of events.

It is possible to somewhat predict the weather for planet Earth with months and years ahead.

In the winter it gets cold, and in the summer it gets warm. Knowing these cycles in nature we can predict the future. Will it not get cold in the next winter?

Based on present events, men can predict the economy, politics, science, and a little bit of everything.

The chaos is comprehensible and can be predicted within probabilities.

When it is winter in the southern hemisphere it is summer in the northern hemisphere, and when it is summer in the southern hemisphere it is winter in the northern hemisphere. Nature has a natural order.

It is not possible, however, to be completely sure about the weather in the future because the chaos changes things quite significantly overtime, deviating from our predictions. It is necessary to find the probability of having this or that weather.

The universe is all rational.

We need to use reason to understand the universe.

If there was not an order in the chaos, nothing would come into existence.

Everything exists thanks to an order and a law. It is this order and this law that scientists attempt to understand.

The most valuable thing about a scripture or an argument is its sense.

Scientists look for a sense in everything.

In fact there is a sense for everything.

Everything can be explained in a simple form.

It is knowledge to simplify and not complicate when it comes to writing or speaking.

All scriptures and arguments need a common sense so that it can become valuable to all and even to the layman.

Even parables need a sense because without a common sense it becomes incapable to teach and pass any valuable information.

Simple words are gold in literature for a gold understanding.

To write and speak in parable is valid only if people are capable of understanding.

If even a child can understand it, then it is because it has been well explained.

It is from good explanations that there is a perfect understanding.

The universe is like an immense clock made of many wheels.

It is possible through the wheels of the universe to foresee the destiny of the cosmos.

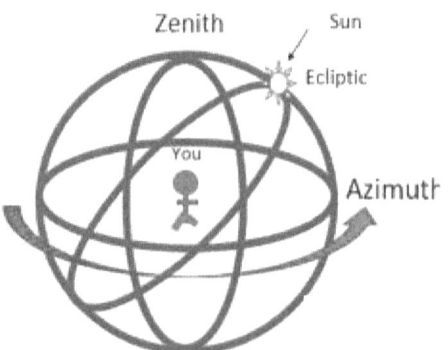

Justice

The important is not the amount known but the attempt to know more, and elaborate what is already known.

It is never too late to learn. It is always time for changes. When we study something hard, a simple understanding is enough to shine some light.

For knowledge, everything is easy and nothing is hard. Everyone who applies the correct effort can learn.

Little by little it is possible to grow intellectually. It is not something that you attain overnight.

It is a human mission to reach perfection. With much will for change we can reach perfection in all of our actions.

Since it is possible to reach perfection we should work on it. Perfection is enlightenment.

If we all act to help the world, we can eliminate all social and environment issues. When we fix all problems, this civilization will be enlightened. This is our mission that must be accomplished for our glory as intelligent beings.

For every action there is a reaction.

Anyone who tries to help the world will profit from it. (Cause and Effect)

All help is needed and it is good. The union of many people adds to a strong force towards a common goal.

For everything there is equilibrium.

There is the right amount of this and that.

If there is a problem it is because there is disequilibrium.

The machine of nature functions in equilibrium.

Equilibrium exists in chemical reactions.

In nature, energy is conserved by being transferred into many forms.

People need to find the equilibrium in their relationship with each other and
nature.

To be in equilibrium, humanity should not allow hunger, misery, no access
to education, health care, and that animals become extinct, or that we destroy the
world through greed, and human actions.

Having fixed all problems, only then will humanity reach equilibrium, the
enlightenment.

Reducing poverty we diminish violence, diseases, and all forms of discontent. It is necessary that many people work together to fight for peace.

Justice is a foundation with norms of what is right and wrong.

For a society to be just it is necessary correct laws.

The laws of justice constitute on what is the most correct in nature.

Each human being has to follow these laws to be considered civilized.

It is the duty of governments to maintain these laws sustaining order.

The politics, art, architecture, literature, music, theater, movies, sports, games, fashion, gastronomy, economy, and science define culture.

Nations need to develop a civilized culture based on the perfection rational men can achieve.

All cultural intentions should focus in the social and environmental benefit.

All actions in the society should respect everyone and the environment.

Animals should be protected against extinction.

Men have the obligation to preserve nature.

The civilization must act in equilibrium, since disequilibrium threatens human life, and all other forms of life.

The life in the material universe is not eternal, but it is human duty to respect life by living in equilibrium.

Geometric Harmony

Everything in nature is immersed in geometric harmony.

This organization favors not only the formation of inorganic matter but also the origin of living organisms.

The Earth is the only planet known to have life, and this makes Earth a very special place. The center of our universe is the Earth.

The probability of having life outside the Earth is small, but exists.

Until now scientists have not found life outside Earth.

Being the only place with known life, Earth is of extreme importance for science.

On Earth there are very complex organic structures that prove that something really out of common happened in this planet.

The organic matter has origin from the inorganic matter.

To understand organic matter it is necessary to study the inorganic matter.

To understand inorganic matter it is needed to study mathematics and geometry that are the two subjects that are the foundation of science.

Mathematics and geometry in the cosmos gave form to inorganic matter which gave structure to the organic matter.

Mathematics is in the foundation of geometry, and the geometric harmony in nature reigns everything.

The cosmic matter is composed of atoms and particles.

The living organisms have one or many cells which are organic matter.

As numerous as there are stars in a galaxy are the numbers of cells in the human body.

Plants and living beings all have many cells in an organic system that is very organized with each cell having many atoms.

A bacterium is an example of an organism with only one cell.

All bacteria, plants, and living beings possess a function in nature. Small and large structures work together in the cosmos.

Being all of nature so organized, nothing is by chance.

We must respect the limits and the equilibrium in nature so that the cosmos can function in way that is beneficial for the health of all living beings.

Diseases have origin from the cosmic disequilibrium.

Life is special and deserves attention and respect.

Of all the stars in our galaxy only one has a planet with life that we know of.

Life exists on Earth from a miracle.

Uflonix finished reading the book with his right index finger up and said:
Uflonix: I guess it worked!
Zeno: What worked?

Then all of the sudden Zeno felt like a helmet was being removed from his head and he faced the desert nomad that he encountered last time.
Nomad: I saw you sleeping in the desert on your way to the ziggurat at Uruk and I decided to use you for my experiment.
Next to the nomad was Uflonix who had joined the nomad in this adventure.
Zeno: Who are you?

Nomad: My name is Jucrilam and I am excited with the results of this experiment. I hope you learned something from this adventure.

Zeno was completely confused, He got up and said: "Indeed these simulations take me far into the farthest reaches of the universe in my mind…..the Ultimate Knowledge ruler of all things….Uflonix you too is involved with this? "

Uflonix: You have no clue. It was my idea to tell Jucrilam to take you in this journey.

Jucrilam: In fact do you remember the explosion?

Zeno: Yes. Do you mean there was never one?

Uflonix: The explosion in the simulation was at the moment you collapsed and hit your head on the ground and then we saw you and immediately placed the helmet over your head.

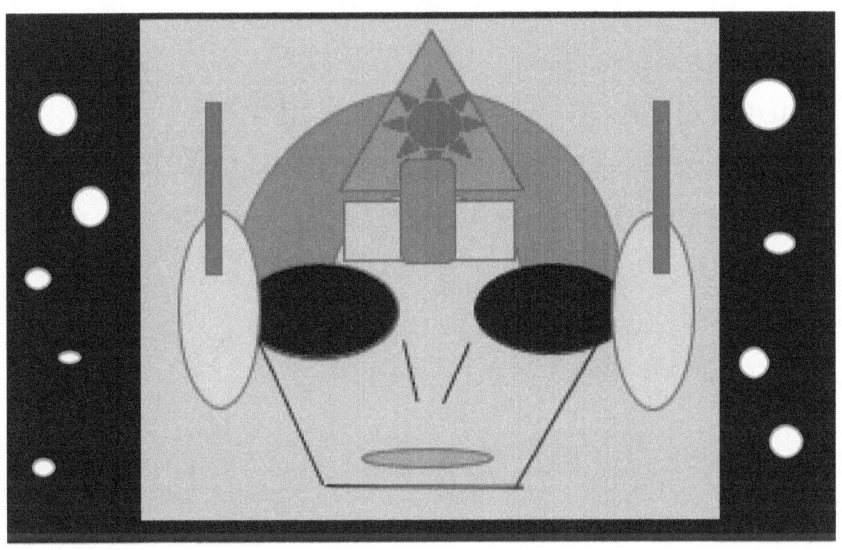

Zeno: Was there ever an explosion? Was there ever a spaceship coming from the sky? Jucrilam laughed and said:

Jucrilam: No there was not. You hit your head on the ground and when we placed a helmet over your head that is when the event became a story in the simulation you were just in.

Zeno looked at both and said: "Indeed I think I need a break from all of this."

Jucrilam and Uflonix laughed and agreed with their heads.

Uflonix: Do take a break but always remember where this helmet took you to. One day you will realize your greatest dream.

Zeno: What dream?

They never answered and simply disappeared from sight like magic leading Zeno to question whether he was still in a simulation or not.

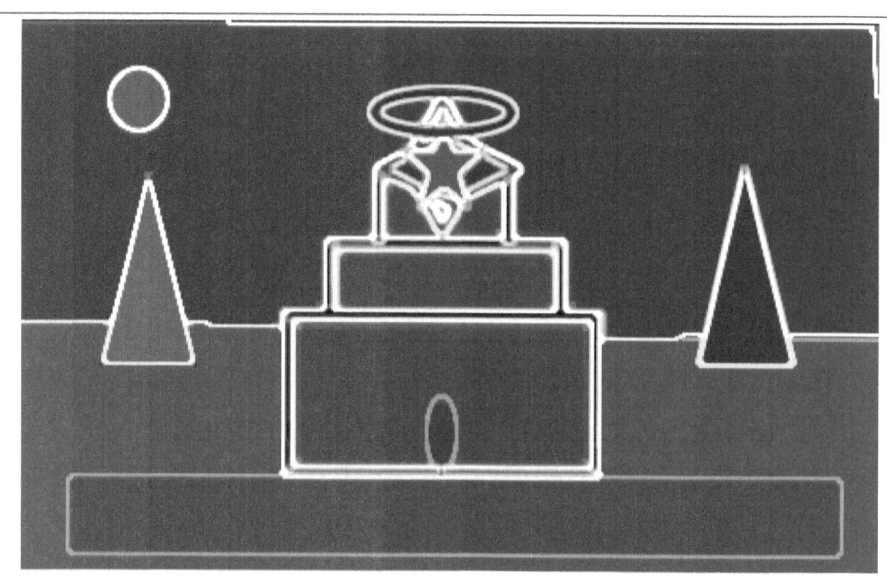

Physics for Those of Some Knowledge

© 2021 Diogo de Souza
All Rights Reserved.

Contact Information:
diogodesouza7@gmail.com
diogodesouza7@hotmail.com

PINDORAMA

Diogo's Universe

Written Diogo Franklin de Souza

© 2017 Diogo de Souza
All Rights Reserved.

The first attempt to this theory began in 2005 while I was a student at a high school physics class. Over time a few modifications have been made but the overall idea and content did not change. 2018

I dedicate this book to the great nation of Israel.

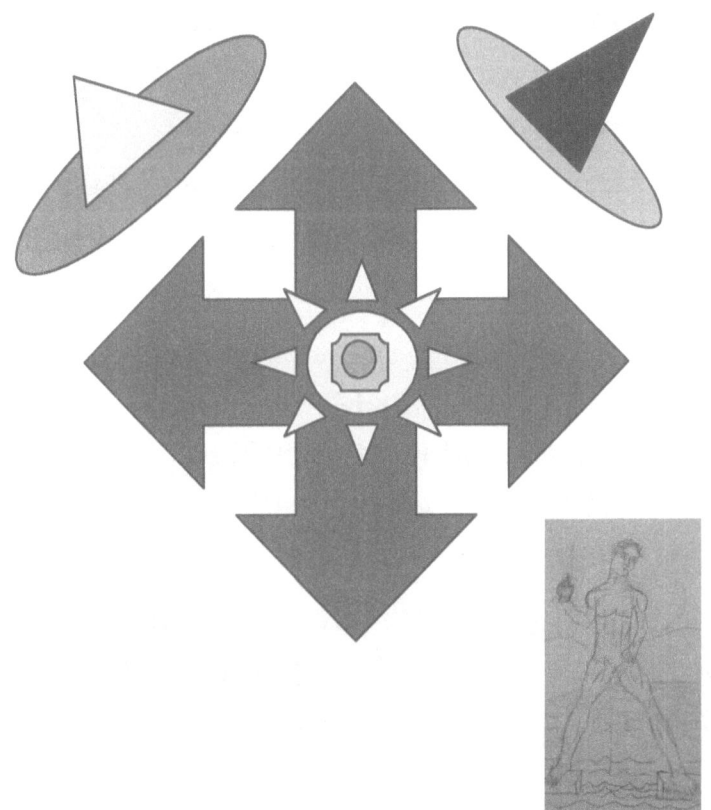

Introduction:

The origin of the matter in the universe are the many quantum fluctuations found throughout infinity giving birth to many other universes. There is no such thing as nothingness. Even in a vacuum, particles are created and destroyed several times. The matter in the universe in the like manner came from these fluctuations that will one day lead to their destruction in the Big Crunch.

Dark Matter is a combination of Rogue Planets, Brown Dwarfs, Higgs Bosons, gases, and Weakly Interacting Particles.

Without Dark Matter space would have no galaxies, stars, and planets. Dark Matter forms a structure in the universe allowing it to stay in place and exist. Dark Matter is also in great part, the Ether physicists referred to in the past.

Matter and Anti-Matter are created from nothing and destroyed after colliding with each other. All Matter in the universe was created likewise, and will eventually collide with the Anti-Matter in the Big Crunch. A new cycle will begin after that.

Dark Energy causes the expansion. Matter and Anti-Matter separate. Later the Dark Energy will cause the cosmos to contract leading Matter to collide with Anti-Matter, and that will be the end of a cycle.

The sum of all of the energy Matter plus Anti-Matter is exactly zero. $1 - 1 = 0$. From nothing, everything came to be.

The average density of the universe at the Planck Level is zero.

The universal sphere contains a doughnut inside with matter travelling along its sides an infinite number of times. There is an infinite number of cycles of creation and destruction.

My Theory in Details

The universe is in the shape of a doughnut inside a sphere.

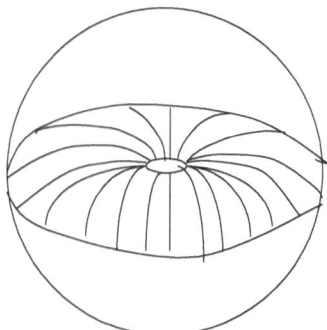

By cutting the doughnut in half and looking through one side we see two wheels. The space between the two circles is the Plank Length.

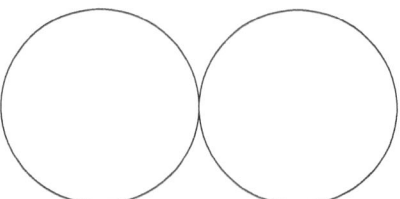

Cutting the doughnut in half as shown.

Imagine cutting the doughnut in half like as follow:
But in my theory the space between the two circles is almost zero.

In the center between the two wheels is the singularity. When the singularity containing all the matter in the cosmic bubble exploded, the expansion of that bubble began. The Matter and Anti-Matter in the universe expand separating from each other due to the Dark Energy caused by a gravitational dipole. $\bar{h} + R - R\cos(0) = \bar{h}$ is the distance between the tip of the arrows.

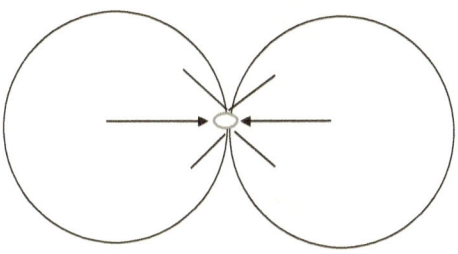

The Planck's Constant above is for length. It is the smallest length possible for the universe. It is how wide was the singularity before the Big Bang.

The arrows indicate position.

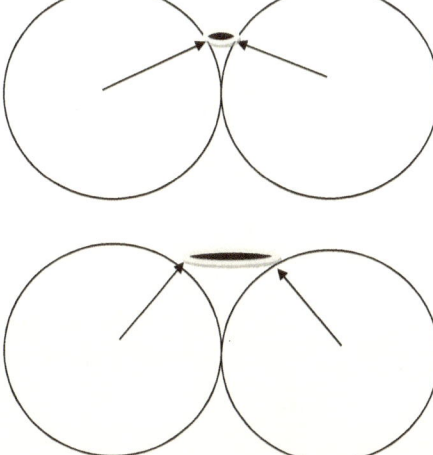

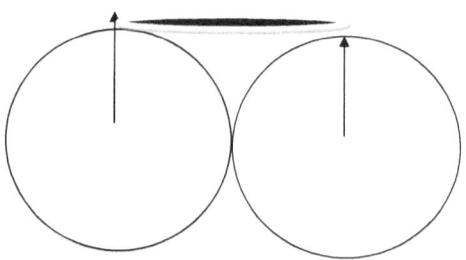

$\bar{h} + R - R\cos(90) = R$

R = distance between the tip of the arrows.

The bubble contains all the matter in the cosmos such as galaxies, and so forth. After stretching along the sides of the doughnut, a hole was formed at the center of the bubble. The hole kept getting bigger until the bubble was transformed into a thin ring.

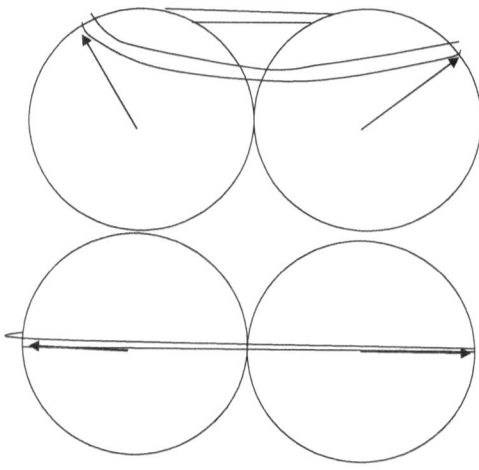

$\bar{h} + R - R\cos(180) = 2R$

2R = distance between the tip of the arrows.

Following the curved path along the sides of the doughnut the cosmic bubble began to contract.

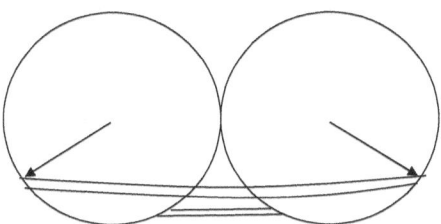

On its way back to the singularity, the cosmic bubble becomes a disk once again, losing its hole in the middle. $\bar{h}+ R - R\cos(270) = R$ is the distance between the tip of the arrows.

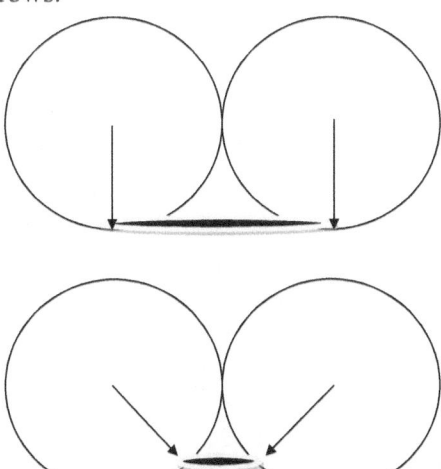

Half of the universal disc is made of matter and the other is anti-matter. When the two separate the universe is created from the singularity, and when the two meet again they are converted to light energy from where a new cycle begins.

The Matter and Anti-Matter in the universe meet again at the Big Crunch. After that, a new cycle begins.

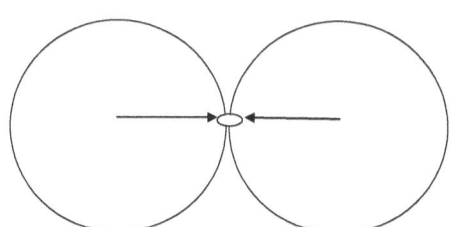
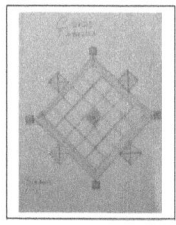

Eventually the cosmic bubble returns to the singularity, exploding in the Big Crunch, becoming a new Big Bang, and doing it all over again and again endlessly.

The equation that describes the distance between the tip of the arrows is = Distance = $\bar{h} + R - R \cos(\bar{h}(z))$

The angle is measured from the singularity to the tip of the arrows and R is the diameter of one of the two wheels.

Plotting the graph of that function we get.

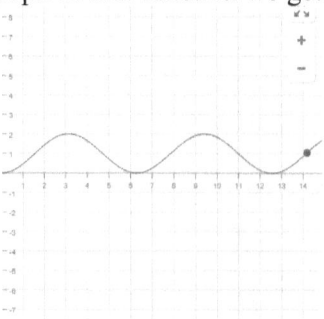

z is angular increments since even angles exist at quantum pieces. z can be 0, 1,2,3,4... and so forth. Only whole numbers which produce the θ needed.

Vertical axis is for distance and the horizontal axis for the angle in radians. When the angle is 180 degrees ($\pi\ radians$), the distance is at its greatest at the peak of the graph. The graph shows about 2.5 cycles of expansion and contraction. The maximum length of the diameter is 2 times the radius of the sphere containing the doughnut.

The total energy of the universe is still zero.
1-1 = 0. This cycle repeats forever since forever.

Particles are vibrating strings shaped like a ring. Particles vibrate Dark Matter (the Ether). That is why particles diffract and propagate in space like waves through a medium. The medium in this case is the Ether.

Everything is made of vibrating rings of energy propagating in space, transferring energy to other rings of energy in the cosmos.

The superposition of these waves forms the cosmos. There is an infinite amount of parallel universes. The closer parallel universes are more similar to ours. The farther parallel universes become more and more different the farther you go.

Travelling back in time is moving to a parallel universe. If you go back to 1965, you will not recognize it. It will not be like the 1965 you learned about. It will look like the year 1965 in a parallel universe.

It is this the reality regarding Feynman's Alternate Histories. There is an infinite number of histories and none of them are exactly alike. There is an infinite number of paths in history both past, present, and future.

Everything in the universe is made of Dark Matter. Visible matter is compressed Dark Matter, and empty space is stretched Dark Matter. Dark Matter is made of strings. When the strings are compressed, waves form giving birth to visible matter. When the strings are stretched, no waves form, and Dark Matter becomes invisible.

The doughnut is curved Dark Matter, and the cosmic bubble is the location of all we can see around us. The curved Dark Matter is the result of a gravitational dipole that pushes the bubble in an endless path around and around again.

This theory solves the problem of a universe coming from nothing, and it instead states that there has always been a bubble following the curved path of doughnut without true beginning and with no true end.

Dark Matter is made of Higgs Bosons that is the Ether. Light needs Higgs Bosons to propagate. Although light has no mass it follows the curved path of space time because it propagates through space using the Ether which are Higgs Bosons.

Higgs Bosons also give mass to particles in compressed regions of Dark Matter that becomes the visible and detected matter and energy that we know.

If light were to not require a medium to propagate it would never follow the curved paths of space time but would go straight instead, but that does not happen, so it does need a medium to flow. Space is not empty but is filled with Higgs Bosons that curve space time. Light travels faster in empty space where Dark Matter is stretched instead of compressed.

On the top half of the doughnut, Dark Energy causes Matter and Anti-Matter to expand, and in the bottom half Dark Energy causes Matter and Anti-Matter to attract. This is called a gravitational dipole.

Proving my theory of the gravitational dipole

1. If heavy black holes at the center of galaxies are also surrounded by a doughnut shaped structure where Matter and Anti-Matter are forced to move as described by my theory, then we could say a similar structure could be happening at the center of universe.
2. If the universe is expanding with positive acceleration but **the change in acceleration is negative**. This means that although the universe is expanding ever faster, the negative change in acceleration will cause it to contract in the future along the curved nature of the dipole.
3. If it is proven that heavy black holes are gravitational dipoles. Same could be said for the black hole at the center of the universe.

If a measurement made today of the acceleration of expansion of the universe, and another measurement taken **1000 years from now**, shows that the change in acceleration is negative, then it means that the universe will contract in the future.

Space can't be smaller than the Planck Length which is $1.6 \times 10^{-35} m$. The singularity had a diameter equal to that number. That is why scientists can't unify gravity with the other forces. There simply is not a way for space to be smaller than the Planck Length. The singularity before the Big Bang was at the smallest possible size. The Planck Length.

Proving my theory of Dark Matter

1 If the path of light curves due to gravity, then it is because it is using the curved medium to propagate. This medium is Higgs Boson, and Higgs Boson is Dark Matter.

2 If light did not use the curved space medium to propagate it would move in a straight line without interacting with the curved structure in space at all.

Let us assume that the arrows of my doughnut shaped universe move around the two wheels (2D model) at a constant angular velocity. These arrows as showed earlier, describe the motion of the edge of the cosmic bubble in its curved path around the doughnut. We get the following:

Distance between the two arrows = $\bar{h}+R - R\cos(\bar{h}(z))$
Where R is the diameter of one of the two spheres. I used R instead of D because R is the radius of the big sphere containing the doughnut. θ is the angle between the arrows and the singularity.

The derivative of the distance between the two arrows with respect to the angle is:

$\frac{dD}{d\theta} = R\sin(\theta)$, this equation gives the change in the distance between the arrows with respect to the change in the angle θ. It is the velocity of either expansion or contraction.

The derivative of that gives the acceleration of expansion or contraction.

$\frac{\partial^2 D}{\partial \theta^2} = R\cos(\theta)$ Change in acceleration per time = $-R\sin(\theta)$

If under observation it is proven that the change in acceleration of expansion is negative, then the cosmos will likely contract and collapse in the big crunch.

$R^2 = v^2 + a^2$
Knowing the velocity and acceleration of the expansion universe, R can be found.

$(R\sin\theta)^2 + (R\cos\theta)^2 = v^2 + a^2$
That is since $v = \frac{dD}{d\theta}$ and $a = \frac{\partial^2 D}{\partial \theta^2}$
V is the velocity and a the acceleration

Using conservation of energy and assuming that energy is homogeneous throughout the cosmic bubble we can concentrate on Planck's Space.

$$\varrho = \frac{1}{2}m(r\omega)^2 + \frac{GmM}{r+} - \frac{GmM}{r-}$$

Where ϱ is the average energy density of a space the size of Planck's Space. This average density is zero. m is the small mass in that space, G is the gravitational constant, and M is the equivalent mass of the black and white hole in the center of the doughnut. r+ is the curved distance between that mass at the very edge of the cosmic bubble from the white hole in the singularity. r- is the curved distance of the very edge of the cosmic bubble from the black hole in the singularity.

A white hole spits matter pushing matter around the doughnut. On the bottom of the doughnut there is a black hole swallowing matter. Both the black and the white hole are in the singularity contributing to the gravitational dipole.

The r in the $\frac{1}{2}m(r\omega)^2$ is the radius of a wheel in the 2D model of the doughnut.

$$\varrho = \frac{1}{2}mr^2\left(\frac{d\theta}{dt}\right)^2 + \frac{GmM}{r+} - \frac{GmM}{r-}$$

Knowing that the average density is zero:

$$0 = \frac{1}{2}mr^2\left(\frac{d\theta}{dt}\right)^2 + \frac{GmM}{r+} - \frac{GmM}{r-}$$

$$0 - \frac{GmM}{r+} + \frac{GmM}{r-} = \frac{1}{2}mr^2\left(\frac{d\theta}{dt}\right)^2$$

$$\frac{2\left(0 - \frac{GmM}{r+} + \frac{GmM}{r-}\right)}{m} = r^2\left(\frac{d\theta}{dt}\right)^2$$

$$\sqrt{\frac{2\left(0 - \frac{GmM}{r+} + \frac{GmM}{r-}\right)}{m}} = r\frac{d\theta}{dt}$$

$$\sqrt{\frac{2\left(0 - \frac{GmM}{r+} + \frac{GmM}{r-}\right)}{m}} = \frac{dx}{dt}$$

$$\sqrt{\frac{m}{2\left(0 - \frac{GmM}{r+} + \frac{GmM}{r-}\right)}} = \frac{dt}{dx}$$

$$dx\sqrt{\frac{m}{2\left(0 - \frac{GmM}{r+} + \frac{GmM}{r-}\right)}} = dt$$

(r+) + (r-) = constant. The distance between a point at the edge of the cosmic bubble and the black hole plus the distance between the same point to the white hole is just the circumference of a wheel in the 2D model of the doughnut.

Let's call the circumference C.

(r+) – C = -(r-)   then   C - (r+) = (r-)

So,

$$dx \sqrt{\frac{m}{2(0-\frac{GmM}{r+}+\frac{GmM}{C-(r+)})}} = dt$$

$$\int_{Planck's\ Length}^{Today's\ length\ for\ r+} dx \sqrt{\frac{m}{2(0-\frac{GmM}{r+}+\frac{GmM}{C-(r+)})}} = T = 13.5 \text{ billion of years}$$

From the above equation we can solve for the (r+) of today's cosmos and verify my gravitational dipole theory. We do know that the current value for (r+) is 13 billion light years away, which is $1.23 \times 10^{26} m$. We also know that 13.5 billion years is 4.09968×10^{17} s. Plugging these values in the above equation we get:

$$\int_{1.6\times 10^{-35} m}^{1.23\times 10^{26} m} dx \sqrt{\frac{m}{2(0-\frac{GmM}{r+}+\frac{GmM}{C-(r+)})}} = T = 4.09968 \times 10^{17} \text{ s}$$

The variable in the above integral is r+. dx stands for the variable r+. The integral goes from the Planck Length at the beginning of this universal cycle, the Singularity at the Big Bang, to the current value for r+. The integral should equal the current age of the universe in seconds.

Little m is Planck's Mass which is 1.78×10^{-41} kg and big M is the mass of the Black-White Hole in the center of the universe which is the total mass of the universe. There are 3 unknowns in the equation which are the value for C and the mass of that Black-White Hole in the center of the universe. If we know C we can then find r-, or either way. $C = 2\pi r$ with r being the radius of one of the circles in the 2D model of the Universe.

Volume of the Cosmic Disk

As the disk shaped cosmic bubble moves along the curved path of the doughnut, its disk becomes thinner and later turning into a thin ring at its maximum diameter expansion. Then it starts getting thicker but the volume decreases until it reaches the singularity with no volume. We can express the thickness of the disk turning into a ring and then into a disk with the equation: $1-1\sin(\theta/2)$

When the arrows representing the path of the cosmic bubble are at 0 degrees, the $1-1\sin(0) = 1$. This means that the disk has no hole in the middle. When the arrows are at 45 degrees, $1-1\sin(45/2) = 0.6173$, which means that 61.73 % of the volume of the disk is made of the cosmic bubble while the remaining 38.27 % is the hole in the middle. The ring keeps getting thinner. When the angle is 180 degrees, 1-

$1\sin(180/2) = 0$, which means that the disk has turned into a thin ring, the thinner it can ever possibly be. This is the only way it can possibly move along the path of the doughnut. The value 0 might seem very radical in saying that the ring is so thin that there is not a ring. In real world that is impossible, but it serves as a mathematical tool to describe a motion of the ring around the doughnut. The Planck Length of space is the minimum it can ever possibly be. We can say that the universe at 180 degrees is as thick as its diameter at the singularity, which is the Planck Length of space (very small). Obviously such an extreme event is really far in the future to happen. We can use the above equation and combine with $R-R\cos(\theta)$ to find the volume of the cosmic bubble as it moves around the doughnut.

Let's say that the disk has a vertical thickness of a fixed number such as Γ, the symbol representing that constant. By vertical I mean by measuring the thickness using a ruler normal to the plane of the disk. The other thickness stands for a ruler measuring the thickness of the bubble parallel to the plane.

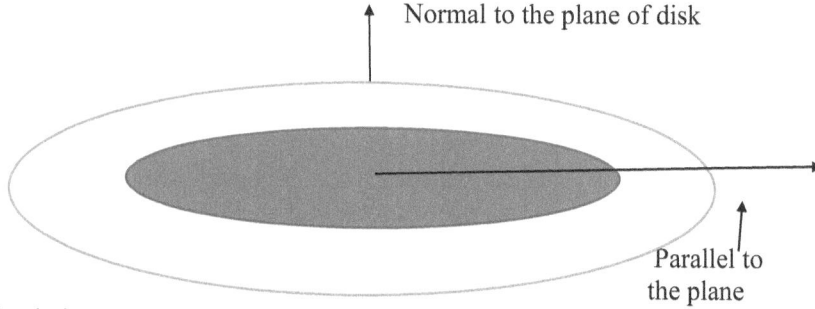

The dark section represents a hole in the middle of the disk.

Let's say that the vertical thickness is constant, but the parallel thickness changes as the disk gets thinner into a ring, or thicker into a fatter ring, or if it has no hole in the middle at all such as in the singularity. The parallel thickness is measured from where the hole begins all the way to the edge of the cosmic bubble ring. A fatter ring is thicker while a thinner ring is thinner.

The area of a circle is πr^2. We know that the diameter of the disk at any given point is R-Rcos(θ) *and that the* radius of the disc is half the diameter so we get: $\frac{R-R\cos(\theta)}{2}$ = radius.

Given the fact that 1-1sin(θ/2) gives the ratio of the region of the disk made up of the bubble compared to the whole disk, and the fact that volume equals areas times thickness we get the following equation for the volume of the cosmic bubble:

(1-1sin(θ/2)) $\Gamma\pi(\frac{R-R\cos(\theta)}{2})$^2 = Volume.

2r = R = Radius of the Doughnut Shaped Universe.

When graphed we get the following in radians:

Γ can be calculated by using the idea that:

$\frac{d(volume)}{dt}$ = Velocity of expansion

$\frac{d^2(volume)}{(dt)^2}$ = Acceleration of expansion

Using the observation of the current universe, its velocity of expansion can be measured and Γ discovered.

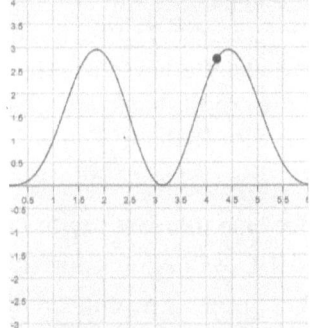

It can be possible to find the θ representing a given point in the universal cycle from the volume of the universe. If we know the volume of the present universe we can speculate what θ is by proposing a value for Γ.

r can be calculated from the equation on the left.

$r = \frac{(r+)}{\theta}$

The vertical axis gives the volume of the cosmic bubble and the horizontal axis the angle in radians. The vertical scale was chosen randomly with the sole purpose to only show changes in volume in the cycle.

The above graph shows 1 cosmic cycle around the doughnut. This shows that although the disk gets thinner, the volume increases until about

1.847radians (105.82 degrees), after that it starts to decrease until the ring disk becomes so thin that it disappears completely in the quantum realm. This happens at around 180 degrees (3.14 radians). Although the diameter of the cosmic bubble disk gets larger, its parallel thickness gets smaller, which shows that it will eventually cause the volume to decrease to the same value that it has in the singularity. Since in today's universe we see the cosmos expanding, most likely we live between 0 and 105.82 degrees in the cosmic cycle. After 180 degrees, the cosmic bubble begins to contract but the parallel thickness starts to grow, turning the ring into a disk. This thickness causes the volume to increase although the cosmos is contracting. This is shown by the second rise leading to a second peak on the graph. The second peak occurs at around 253.71 degrees (4.428 radians). Then the contraction of the cosmic bubble, despite the growing parallel thickness, leads to a decrease in volume until reaching the singularity with 0 volume in the quantum realm at 360 degrees ($2\pi\ radians$). **In the equation 1-1sin(θ/2), only degrees between 0 and 360 are allowed or else the graph and data stops making sense as the graph shoots for angles greater than 360 degrees when combined with R-Rcos(θ) with no physical meaning for it.**

The entropy in the universe increases with volume between arrow angles of 0 and 105.82 degrees and decrease with volume between that and 180 degrees. The entropy then increases again with volume between 180 degrees 253.71 degrees to then decrease with volume from that to 360 degrees. One complete revolution of the cosmic bubble around the doughnut gives a change of entropy of exactly zero. The universe is self-contained and fully enclosed.

Temperature:

Treating the cosmic bubble as a gas we can see the relationship:

Temperature = $\dfrac{\sigma}{Volume}$

Temperature = $(\sigma 2)(Pressure)$

The temperature is related to the volume of the gas. The smaller the volume the higher is the temperature. The σ and $\sigma 2$ are constants of proportionality. This relation is the opposite of $\dfrac{PV}{T}$ from the ideal gas law that shows that the smaller the volume the lower the temperature. In our case here, the smaller the volume the greater is the temperature instead. Also in our case, the higher the pressure the higher is the temperature.

So our general equation should be instead:

$T = \frac{P}{V}(\sigma)$ Notice that we are solving for temperature. When the pressure increases, the volume must simultaneous decrease in order for the temperature to get bigger.

The temperature of the universe is the greatest at angles 0 and 180 degrees when its pressure is the highest and its volume the smallest.

The current universe is cooling which shows that the pressure is getting smaller and the volume bigger.

Shape of the Cosmic Bubble:

Bellow is the illustration of the cosmic bubble:

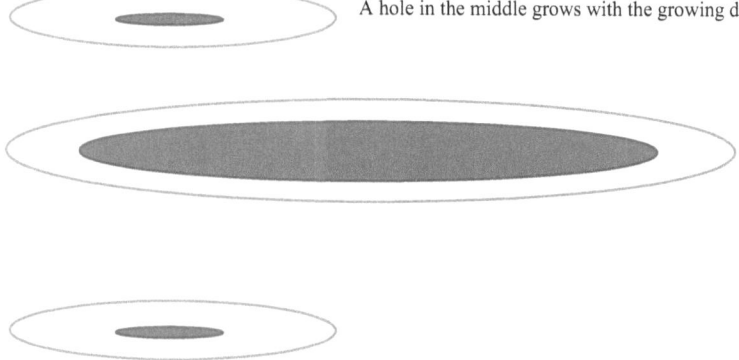

At singularity no hole is present

A hole in the middle grows with the growing disk

Back to Singularity the hole disappears

To prove that the cosmic bubble is a disk we will need to observe a pattern in the cosmic microwave background.

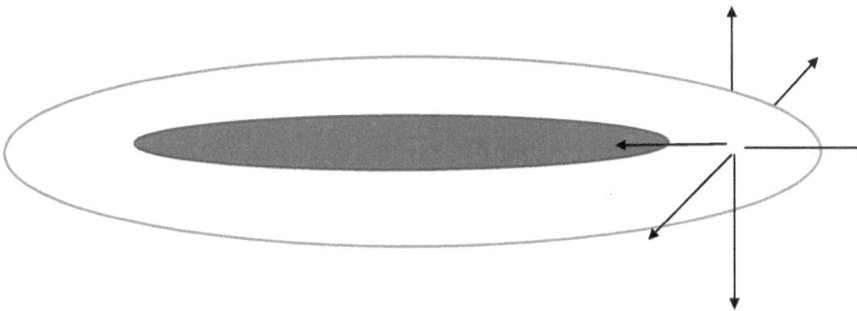

The source of the arrows is a probable location of our galaxy in the cosmic bubble.

From this picture we conclude that because the disk is thinner than it is wide we expect to see more matter in the following directions:

Both locations are 180 degrees from each other. In the cosmic microwave background that would indicate hotter regions.

We would expect the two cooler regions, not as hot as the first case to be in the following directions:

Once again at 180 degrees from each other.

And last but not least, we expect the coolest of all the regions of the cosmos to be the ones looking in the direction of the thickness of the disk:

Also at 180 degrees from each other.

Now we should be able to see a pattern like that in the cosmic microwave background. Since our galaxy and our solar system may be tilted when compared to the plane of the cosmic bubble, what we see will also be tilted. Here is the picture of the cosmic microwave background:

The brighter regions are the hottest.

Cyclical Universe and the Wheel

$$(1-1\sin(\theta/2)) \, \Gamma\pi\left(\frac{R-R\cos(\theta)}{2}\right)^{\wedge}2 = \text{Volume}.$$

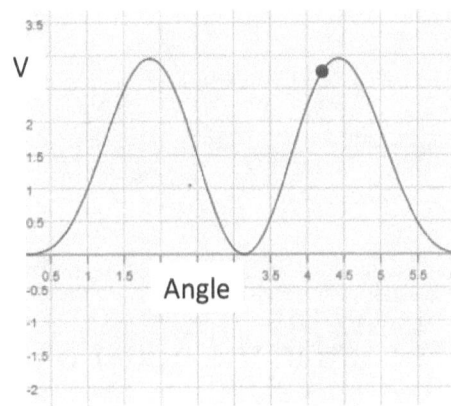

The above equations shows the variation of the Volume of the Universe as the angles in my Dipole Theory changes. The graph represents that function. Γ is the parallel thickness which is held constant.

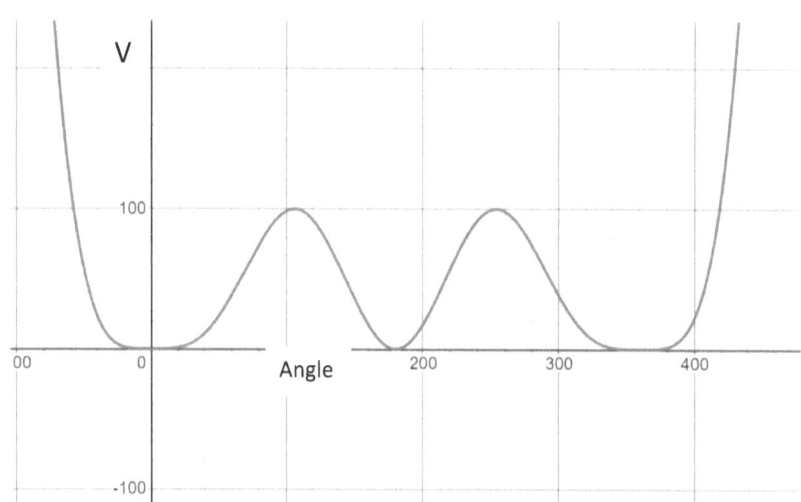

The angles used in the equation on previous page must be between 0 and 360 degrees. Any numbers beyond 360 degrees, or below 0 degrees will cause the graph to shoot up as shown in the graph above. The shooting up has no physical significance.

100 was chosen as the largest volume that the Universe can have as a constant of proportionality with no physics significance.

Angle	Volume	Angle	Volume
0	0		
15	0.30730102	195	9.9568646
30	4.0502649	210	35.968446
45	16.125776	225	67.431335
60	38.072427	240	91.800631
75	65.483323	255	99.887744
90	89.275964	270	89.501078
105	99.868705	285	65.815699
120	92.009366	300	38.388226
135	67.794665	315	16.336273
150	36.344381	330	4.1388495
165	10.193353	345	0.32186402
180	3.868227×10^{-4}	360	7.837042×10^{-9}

Above is the graph for the Volumes with respect to the angles during one complete cycle of oscillation.

Temperature:

$$\frac{(Z)}{(Volume)} = \text{Temperature}$$

When Volume is equal to Planck's Constant Temperature is equal to: 10^9 Kelvins.

$10^9 Kelvins$ = Highest Temperature the universe can have at the Singularity.

$$\frac{(L)}{Volume} = \text{Pressure}$$

Z and L are constants of proportionality.

Since $Volume = \frac{Z}{Temperature}$

$$\text{Pressure} = \frac{L(Temperature)}{Z}$$

L/Z = Y another constant of proportionality.

Linear relationship:

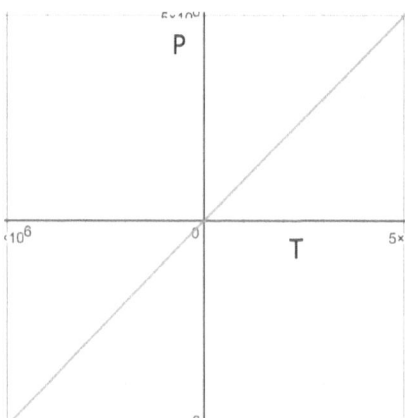

The relationship between Pressure and Temperature is a straight line.

Since $\frac{(Z)}{(Volume)} = \text{Temperature}$

$$\text{Pressure} = \frac{L}{Volume}$$

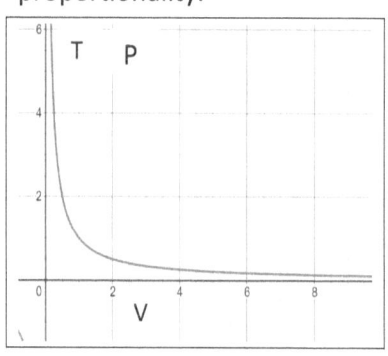

There is an inverse relationship between (Temperature and Volume), and (Pressure and Volume).

Distance between the two arrows = $\bar{h} + R - R \cos(\bar{h}(z))$

z is angular increments since even angles exist a quantum pieces. Z can be 0, 1,2,3,4... and so forth. Only whole numbers which produce the θ needed.

The above equation only gives the variation of the diameter of the ring with respect to the angle.

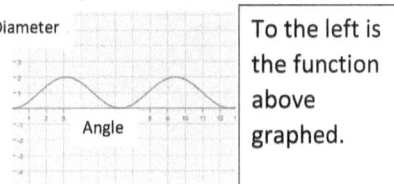

To the left is the function above graphed.

$R^2 = v^2 + a^2$
Knowing the velocity and acceleration of the expansion universe, R can be found.

$(R\sin\theta)^2 + (R\cos\theta)^2 = v^2 + a^2$

That is since $v = \dfrac{dD}{d\theta}$ and $a = \dfrac{\partial^2 D}{\partial \theta^2}$

V is the velocity and a the acceleration

$\dfrac{dD}{d\theta} = R \sin(\theta)$, this equation gives the change in the distance between the arrows with respect to the change in the angle θ. It is the velocity of either expansion or contraction.

The derivative of that gives the acceleration of expansion or contraction.

$\dfrac{\partial^2 D}{\partial \theta^2} = R \cos(\theta)$ Change in acceleration per time = $-R\sin(\theta)$

Step 1

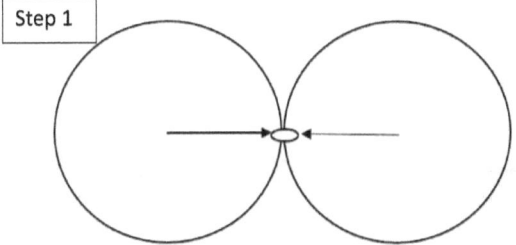
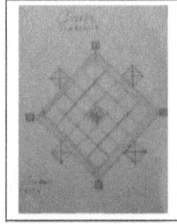

Doughnut Theory in 2 dimensions:

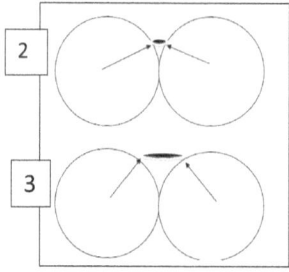

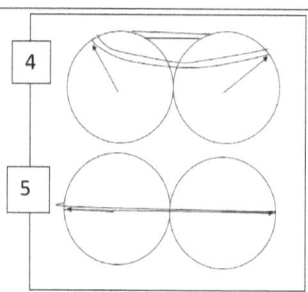

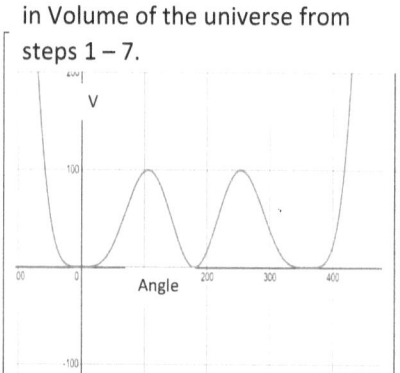

Below is the graph of the change in Volume of the universe from steps 1 – 7.

The Equation below can be used to find r+ which is the distance from the Earth to the Big Bang.

$$\int_{1.6 \times 10^{-35} m}^{1.23 \times 10^{26} m} dx \sqrt{\frac{m}{2(0 - \frac{GmM}{r+} + \frac{GmM}{C-(r+)})}} = T = 4.09968 \times 10^{17} \text{ s}$$

$$\varrho = \frac{1}{2}m(r\omega)^2 + \frac{GmM}{r+} - \frac{GmM}{r-}$$

The equation of the top was derived from the one the left using the fact that ϱ = zero= Total Energy of the Universe.

The Five Elements that make up the Universe according to Plato:

Land Water Air Fire Ether

All the universe is made of Energy. All matter is Energy. All Energy are waves oscillating in the strings that make up the Dark Matter.

Empty space is made of stretched strings:

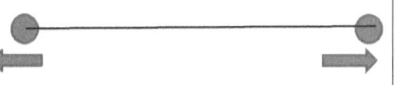

Space containing matter is made of string with waves in it.

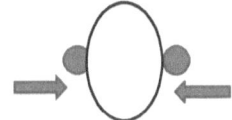

Gravity is caused by matter pulling the strings in the Dark Matter (Ether) leading to it being attracted to other matter.

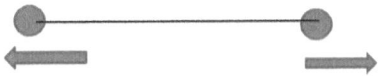

The balls at the end of strings are the Higgs Bosons. When they approach each other gathering, they form a wave in the string between them leading to matter. When the Higgs Bosons move away from each other no waves are formed and that is empty space.

Charges are caused by the amount of rotation of a particle in Dark Matter. This rotation generates a rotating fluid of light in space leading to + or - charges.

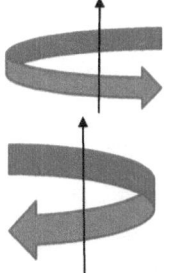

The direction of the spin with respect to a magnetic field generates a charge + or - charge. This spin has nothing to do with the spin in quantum mechanics.

2r = R = Radius of the Doughnut Shaped Universe.

.r = Radius of one wheel in 2D

It can be possible to find the θ representing a given point in the universal cycle from the volume of the universe. If we know the volume of the present universe we can speculate what θ is by proposing a value for Γ.

r can be calculated from the equation on the left.

$$r = \frac{(r+)}{\theta}$$

(r+) + (r-) = constant. The distance between a point at the edge of the cosmic bubble and the black hole plus the distance between the same point to the white hole is just the circumference of a wheel in the 2D model of the doughnut.

Let's call the circumference C.

(r+) − C = −(r-)   then   C − (r+) = (r-)

So,

$$dx \sqrt{\frac{m}{2(0 - \frac{GmM}{r+} + \frac{GmM}{C-(r+)})}} = dt$$

$$\int_{1.6 \times 10^{-35} m}^{1.23 \times 10^{26} m} dx \sqrt{\frac{m}{2(0 - \frac{GmM}{r+} + \frac{GmM}{C-(r+)})}} = T = 4.09968 \times 10^{17} \text{ s}$$

From the above equation we can solve for the (r+) of today's cosmos and verify my gravitational dipole theory. We do know that the current value for (r+) is 13 billion light years away, which is 1.23×10^{26} m. We also know that 13.5 billion years is 4.09968×10^{17} s.

The more compressed a region of space is, with Higgs Bosons clustered in one place, the greater is the Gravitational Pull.

In empty space the Dark Matter (Ether) grid containing the Higgs Boson Spheres has its strings stretched.

In space containing matter, the Higgs Bosons are compressed close together, leading to waves in the Dark Matter String.

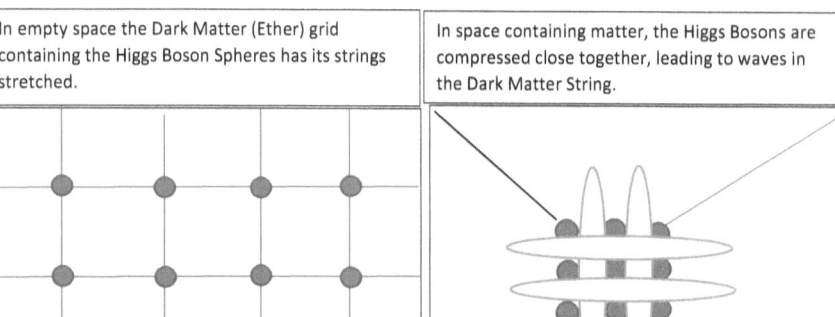

The Four Forces of Nature:

Gravity: Caused by the pull of the strings of Dark Matter.

Electromagnetic, Weak, and Strong: Propagation of light waves in the Dark Matter medium.

Waves: All matter is energy. All energy is light. All light are waves in the Dark Matter medium.

Everything is made by light since everything is energy.

When a particle or any matter travels through space it uses the Dark Matter of the region it is travelling through as part of their composition. Example is shown below.

Here is empty space with Higgs Boson equally separated from each other.

Here now is a particle travelling through that region of space towards the right:

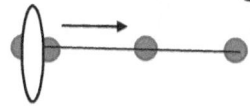

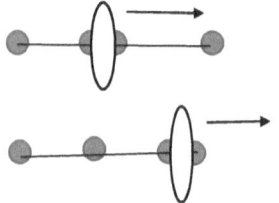

Notice that Dark Matter is used by the particle as its composition. The particle literally goes right through Dark Matter and that is why it is so hard for Dark Matter to be detected.

The strings get compressed in the presence of a particle forming a loop on the particle's location.

Rings of Energy

The 3-dimensional axis:

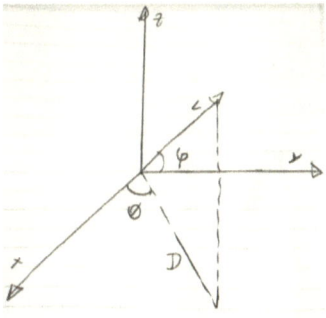

L is the distance between the Particle and the point (0,0,0).

D is the line distance between a point directly below or above the Particle on the x,y plane to the center (0,0,0).

θ is the angle between the x axis and the line D.

φ is the angle between the x,y plane and the Particle.

Using these 3 coordinates, we get the following for each coordinate:

X	Y	Z
$L(\cos(\varphi))(\cos(\theta))$	$L(\cos(\varphi))(\sin(\theta))$	$L(\sin(\varphi))$

These come from the following:

$X^2 + Y^2 = D^2$

$D(\cos(\theta)) = X$
$D(\sin(\theta)) = Y$
$D^2 + Z^2 = L^2$

$L\sin(\varphi) = Z$
$L\cos(\varphi) = D$

$X^2 + Y^2 = L^2 - Z^2$
$X^2 + Y^2 + Z^2 = L^2$

Let us say that the Zyn Particle is moving in a helix pattern in the x direction. Looking at it from above we see the helix and the Particle going in the x direction. Looking at the motion by being in front of it, we only see a circle in the y,z plane:

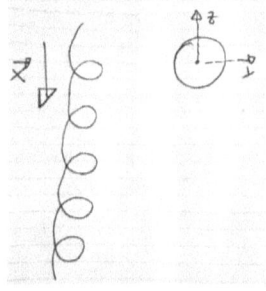

We get the following equation from the coordinates from this motion:

X = (speed)(time)

Y = Lcos(φ)

Z = Lsin(φ)

Since $\theta = 90°$ and sin(90) = 1 and cos(90) = 0

Let us say that the particle rotates in the y,z plane while moving in the x direction at angular Frequency ω_φ. We then get the following:

X = $v_x t$

Y = Lcos($\omega_\varphi t$)

Z = Lsin($\omega_\varphi t$)

Let us say that while the Particle moves in a helix pattern in the x direction, in itself oscillates sinusoidally with Frequency ω_L which means that the value of L changes according to:

L = $L_o \sin(\omega_L t)$

The new equation for Y and Z then becomes:

Y = $L_o \sin(\omega_L t)(\cos(\omega_\varphi t))$

Z = $L_o (\sin(\omega_L t))(\sin(\omega_\varphi t))$

Let us say that this Particle besides moving in a helix in a sinusoidal pattern it is also moving through a tube that forms a perfect circle around a center point. The distance between this center point and the center of the tube is R.

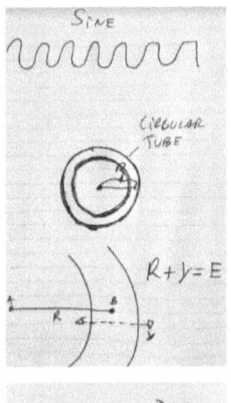

The distance between the center of the circle and the inner edge of the tube is R plus the maximum negative value of y. The distance between the center of the circle and the outer edge of the tube is R plus the maximum positive value of y. The equation for the distance between the center of the circle and the horizontal component of the location of the Particle inside the tube is E = R + Y

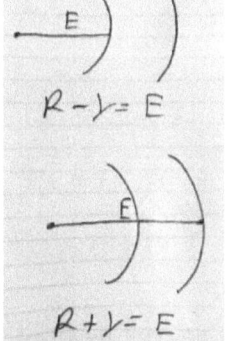

E is the horizontal component of the distance between the Particle and the center of the circle along the x,y plane where E is located.

E = R + Y

The x and y in the x,y plane of E is not the same X and Y of the Particle inside the tube.

Let us say that the Particle moves around the tube with Frequency ω_E

We get the following:

$E_x = E\cos(\xi)$ — With ξ the angle between the line E and the positive x axis of the x,y plane in E.

$E_y = E\sin(\xi)$

$E_x = [R + L_o \sin(\omega_L t)(\cos(\omega_\varphi t))]\cos(\omega_\xi t)$

$E_y = [R + L_o \sin(\omega_L t)(\cos(\omega_\varphi t))]\sin(\omega_\xi t)$

The true distance between the Particle and the center of that circle must also take in consideration its z position so then we have this distance equal to ψ:

$E^2 + Z^2 = \psi^2$

Unlike x and y, the z coordinate of the particle inside the tube is the same z of the particle above and below the plane where E is located.

With

$Z = L_o(\sin(\omega_L t))(\sin(\omega_\varphi t))$

The equation for ψ is then using Pythagorean Theorem:

$$\psi = \sqrt{[R + (L_o \sin(\omega_L t))(\cos(\omega_\varphi t))]^2 + [L_o(\sin(\omega_L t))(\sin(\omega_\varphi t))]^2}$$

Since sine square plus cosine square equals 1 the $\cos(\omega_\xi t)$ and $\sin(\omega_\xi t)$ are gone is the equation above.

If v is equal to the speed of an electron = $2.16 \times 10^6 m/s$ and R the radius of a Hydrogen Atom = $5.29 \times 10^{-11} m$.

It takes the Particle $1.54 \times 10^{-16} s$ to make one revolution.

Which makes its ω_ξ = 4.08x10^{16} radians/second

ω_L = Is the Frequency of the Particle oscillating sinusoidally which is equivalent to its Energy.

ω_φ = is the Frequency of the Particle rotating in a helix. The faster the rotation the more charge the Particle has. Clockwise and Counterclockwise motion leads to + or − charges. No rotation in helix is equivalent to a neutral Particle.

Particles can move at any pathways besides a circular tube:

Three- Leaf Rose

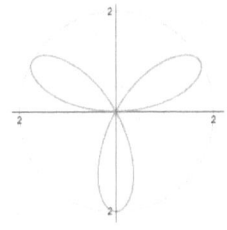

Four-Leaf Rose

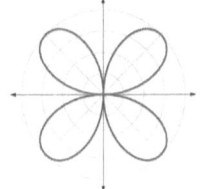

Lemniscate

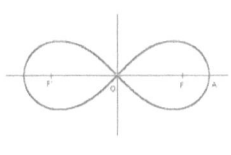

These equations can be very complex and be able to give an in depth investigation of the probabilistic wave cloud of Electrons in orbit of Atoms. **Maybe the motion of Electrons and Particles in general are not at random inside probability clouds, but do follow a repetitive pattern in which it is so complex that we can only grasp it by using probabilities.**

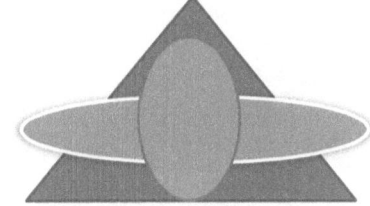

Jucrilam

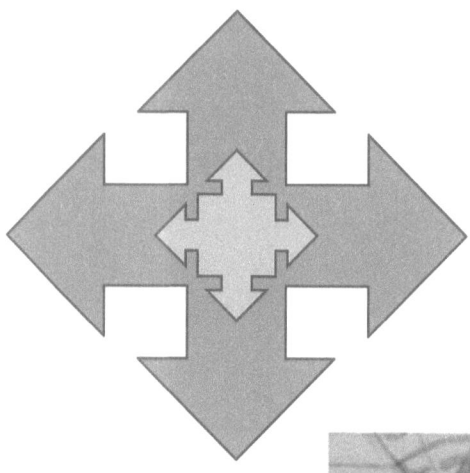

© 2021 Diogo de Souza
All Rights Reserved.

Contact Information:
diogodesouza7@gmail.com
diogodesouza7@hotmail.com

Zeno

Zeno our main character was in search of something great. He could not understand the world around him and thought on the possibility of another world. A world without evil and that is entirely perfect. Would people play games against each other in a perfect world? Would people rather agree that we are all equal and meant to not compete? How would the universe be without death? Zeno walked over the steps along the sea shore in Greece during a sunny day and a nice breeze. He was in the land of Ancient Philosophers and he had many questions unanswered. He decided to walk towards a tree that was in front of an old man wearing all white. The man turned around and bumped into Zeno in an aggressive move.

Zeno: Who are you?

Zeno then looked deep into the man's eyes and fully recognized him.

Zeno: Jucrilam! I found you in Greece!
Jucrilam: No! I followed you instead.
Zeno: What does that mean?

Jucrilam: It means that I am not done with you yet! I have a scroll that will be able to answer some of your questions.

Jucrilam gave Zeno a scroll.

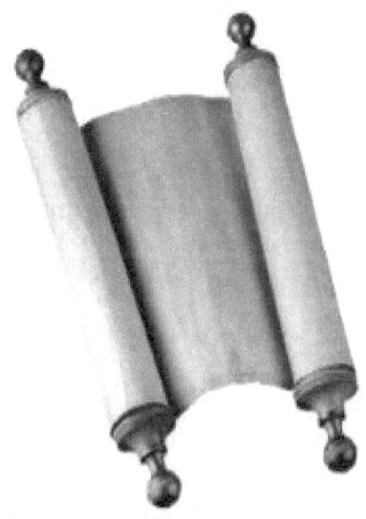

Zeno: What questions are you talking about?

Jucrilam: About worlds with no evil, and where the main goal is the same as that of philosophers. To just contemplate infinity eternally. Here we are under the sun near the shore in Greece. I have a white vest here for you in my bag. Read what is in the scroll and enjoy the sunshine and this nice breeze. Enjoy the life of an ancient Greek Philosopher.

Zeno was then given a white vest in which he placed over the one he was currently wearing and grabbed the scroll curious on what he was going to learn in another one of these strange days of his which were more common than normal days.

Zeno then began to read Jucrilam's comments on two of Aristotle's books:

Investigating Aristotle   pages 6 to 15 ➡

Categories

Humans name things but what are the names? The names are labels given to call things. Without naming them it would be difficult to talk about them. It is a way to distinguish one thing from the other. When using the word animal, that can be a description of a human and a bird. A more in depth description will appear unique for a man distinguishing him from a bird. There are names that are general that refer to an entire class of classification. Both a man and a bird are animals, but each is different. Likewise each man is different from another man. The only reason that there is a word man is because men exist. Without men the word man would not exist since the words are labels given to things, animals, or people that exist. The existence of a man leads to the word man. The realization that a man is an animal leads to the word animal in describing a man.

It is important to describe things, and to name everything so that reason can maneuver its gears while explaining all things around.

There is substance which are individual objects, or living beings or things. There is quantity which is the amount. There is quality which explains what kind, white, black, grammatical and so forth. There are relations such as double, half, greater, smaller, yesterday, tomorrow. There is position such as sitting, armed, state, action.

There are affirmations and negations which is formed from a combination of all of these terms.

The order of classifications are important.

A man belongs to species man, which belongs to the genus animal.

One fact to consider is that the word white can describe a chair, but white does not necessarily indicate a chair. Not all chairs are white. White is something present in a subject but labeling something as white does not lead to the actual subject since subjects can be of many different colors.

Animal

Specie

Man

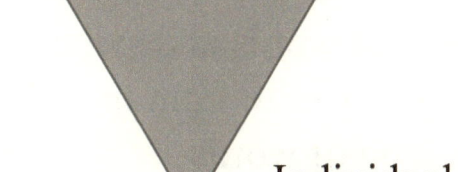

The upside down triangle

Individual

Substances have no contraries such as there is no opposite of man, or opposite of bird. A man is a man and a bird is a bird and there are no opposites. When a substance does change itself it admits contraries only in its qualities. Such as turning white into black, sitting instead of standing up, or doing something bad instead of good.

Quantities can be discrete or continuous in space. Some quantities are positions relative to other parts. Such as being between two things as opposed to being in front of the two things.

The speech is made of discrete quantities, composed of words and syllables. Each syllable is separate from the other in the composition of words.

A line is continuous with points as boundaries while a plane is continuous with lines as boundaries and solids have boundaries that can be points or planes. Space and time are quantities with time being past, present, and future forming a continuum.

Some parts have position while others do not. A line can be parallel to another line, but words that are being said do not have a position since they are only words without place such as a plane, a line, and a point. Words are abstract but they are still a discrete quantity.

Quantities are said to be relative to other quantities. A double is said to be double with reference to something else. A half is said to be a half with reference to something else.

Quality is how and individual is said to be such and such. There is the habit which is long lasting and firmly established, and there is disposition which can be easily changed and quickly gives place to its opposite. Quality can be characteristic of virtue such as being a good or bad boxer, healthy or sick. People are not good due to dispositions but rather because of an inborn capacity to be good.

There is also effective quality of things. These are bitterness, sweetness, sourness. There is also hot, cold, whiteness, blackness. Things contain these qualities but they are not these qualities. A fruit may be sour not because it is sour but because sour is contained in it. An object is called black not because it is black but because it contains a black color.

Substances are affected by consequences. A man blushes when in shame, or turns pale when scared not because he is like that but because that is how he feels at that given moment.

Another form of quality is its shape such as straightness and curviness, rarity, density, roughness, smoothness. It is how the shape of objects are described in reference with other objects.

Interpretations:

Words are symbols of mental experience. Thoughts are expressed in actions and words spoken or written.

The mind is complex. It is a product of the brain but it is not possible to describe thoughts unless there are actions or words spoken or written.

There are affirmations and denials when words are being used. Every affirmation has an opposite denial, and every denial an opposite affirmation.

There are universal things and individual things. Man is universal, while Diogo is an individual.

Universal claims could be contrary since not all things are the same. Individual claims are more specific and more likely correct. In saying that all men are white is wrong, but stating that to a single man then that might be correct.

A pair of contraries can sometimes be both true with reference to the same subject.

"Not every man is white"

"Some men are white"

Is an example where they are both correct.

Affirmations and denials, however, are either true or false.

When future events have alternatives there is potentiality in contrary directions. The affirmation and denial are both possible and have the same characteristic.

There are things that exist in potentiality but not in actuality. Something could be but whether it is or not is not known. A book on a table may fall in the next hour or may not. A person can push on it and it falls or it might stay over the table. The book has the potential to fall but if it will or not is unknown.

Universal statements are usually wrong:

"Every animal is just"

"No animal is just"

Both will never be true. Sometimes the contradictories of the contraries will both be true:

"Not every animal is just"

"Some animals are just"

They are both true.

Universal claims such as:

"Every man is wise"

"Every man is not wise"

Are wrong statements.

A contradictory statement such as:

"Not every man is wise"

Is correct.

The unity of the words are linguistic and not real. Words are not real but rather invented by men and only exist abstractly.

The arrangement of the words gives them unity.

"A man that is white is musical"

White and musical do not always go together but in this sentence they do since they form a unity in the description of the man who is the subject.

"A good man is a shoemaker"

It does not mean that he is a good shoemaker. It just means that he is a shoemaker and is also a good man.

Things that are potential may be in the potential state or in the act state.

A man can walk but may not be walking right now. He may be sitting but still with potential to walk.

The use of the word maybe:

"It may be"

"It may not be"

Could both be correct simultaneously.

The use of the word necessary:

"It is not necessary that is should be" is not the negative of "It is necessary that it should not be". They both mean the same thing.

"Impossible that it should not be"

"It is necessary that it should be"

The word necessary follow the word possibility.

"Necessary that it should be"

"Possible that it should be"

Maybe is twofold with being either it or not it. If one is true the other then becomes false.

There is no logical impossibilities that can occur when there are words arranged in the correct manner and compared with words of another sentence.

Nothing actual is twofold. You can't walk and not walk at the same time. Fire does not heat and cool at the same time.

Potentialities lead to actions. Actions can't occur with its opposite simultaneously.

Eternal is always prior to the universe. Actuality is prior to potentiality. There are actualities without potentialities such as seen in primary substances. There are also things that are always potential but never becomes actuality.

When something is good it is contrary to not good. Not good means that it is bad. That something is good is said not to be good is a false judgement. To further state that it is bad is accidental in nature. So that means that errors are transitions.

Two true statements are never contrary. They are connected and parts of truth.

It is not good is the same as being bad.

No true statements or judgements can be contrary to the other. They are consistent.
Here ends the comments on Aristotle for now. ------------

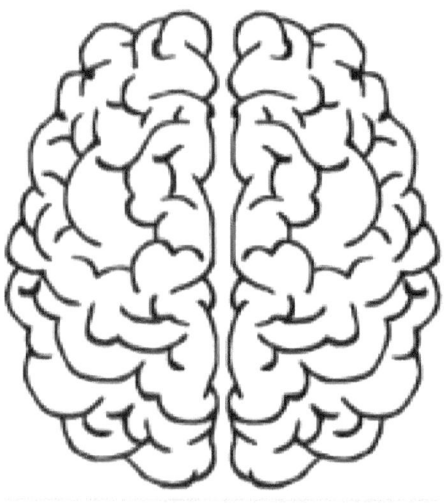

Exploring how humans understand and learn things

Zeno then noticed that the scroll was attached to other pieces of literature parts of different books taped together. Zeno then continued reading unaware of his surroundings and completely focused on his reading.

Law and Order

There are universal rules that keep the order of the cosmos. These are common sense in which anyone of good reason will agree. That means that these rules converge towards a consensus. They are rules to follow which constitute what we call justice so discussed by Socrates in the Plato's Republic. If there is no justice the universe breaks apart since it is what is right and not what is wrong that keeps the constituents of matter together giving shape to the cosmos. Humans are meant to live in a just society ruled by fairness and that fills the community with a sense of wellbeing, and which binds everyone together as a single flock, followers of the great light.

Book of Easiness:

Easiness is capable of doing complicated things easily.

Overthinking is complicating things and is highly unnecessary and it is wrong.

Ideas of consciousness may lead to overthinking on matters that are not worthy. It makes the mind tired and not as capable to perform complicated tasks. Not thinking about consciousness and by just living a life allows us to do difficult things without a sweat. We are able to perform more complex activities while being less tired since we are truly living. This is probably the most important hidden teaching of Christianity.

The parable of King Arthur

Very strong men tried with their strength to remove a sword from a rock. They thought that with their strength they would be able to do it. They relied on physical reasons and believed that only the strongest and most important of all the men would be capable of such task. Then comes Arthur who was a child and filled with innocence removed the sword from the rock with the most act of kindness and simplicity. That is exactly how things actually work. It is not by overcomplicating that we are able to do difficult tasks, but by following easy steps we can perform the most difficult job in a much easier manner. It is not unusual for kids to see better than adults and do things that adults simply can't simply because kids never overthink.

Jucrilam interrupted Zeno and said:

Jucrilam: Aristotle discussed the label of things, the fact that we humans give names to all that exists. We also divide things and put them in order and in different degrees of importance. The universe works in the same way and there was a division between good and evil, between what is and is not, what is good and bad. There must, however, be another reality. One in which the unpleasant is never present. Do you know how that can be?

Zeno: I would seriously like to meet that reality.

Jucrilam: First let me show you star pictures taken by Ogo during his ventures in Texas under light polluted skies.

They then continued reading the scroll.

Binocular Stargazing Targets

Having binoculars is possibly the best way to enjoy stargazing. Here is a list of the best targets from Dallas, and Mineral Wells Texas. Pictures taken from an Android Phone and a 15X70 binoculars. Pictures labeled DFW were taken in Dallas, and pictures labeled MW were taken in Mineral Wells.

In the next pages there are 20 pictures on interesting objects in the sky:

Fig 1: Taurus (MW)

Fig 2: Orion Belt (MW)

Fig 3: Orion Nebula (MW)

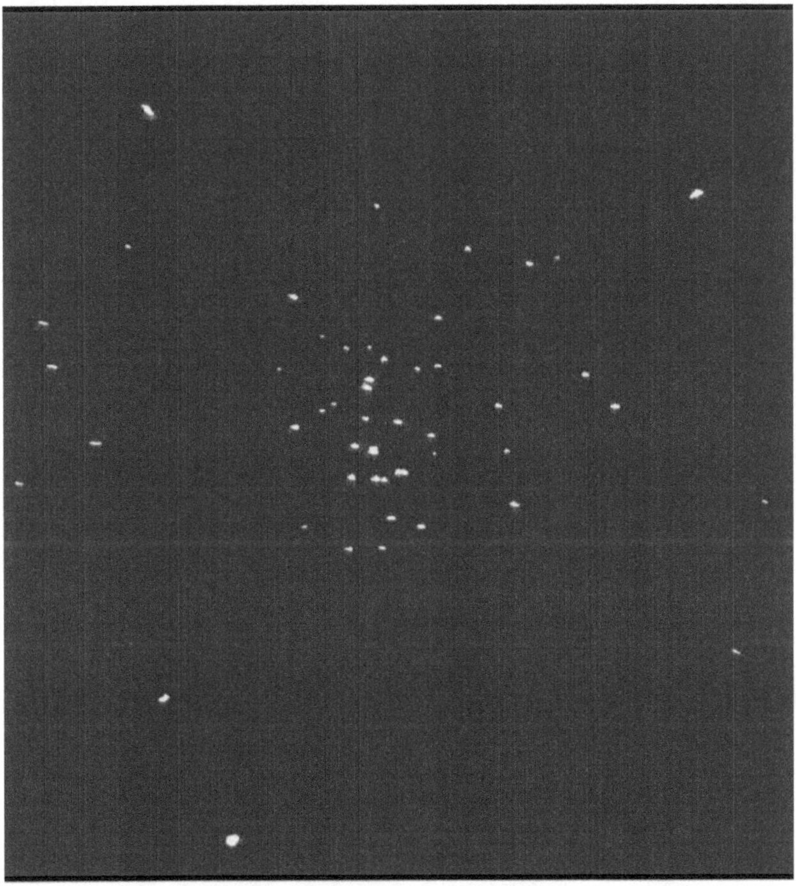

Fig 4: Beehive Cluster (MW)

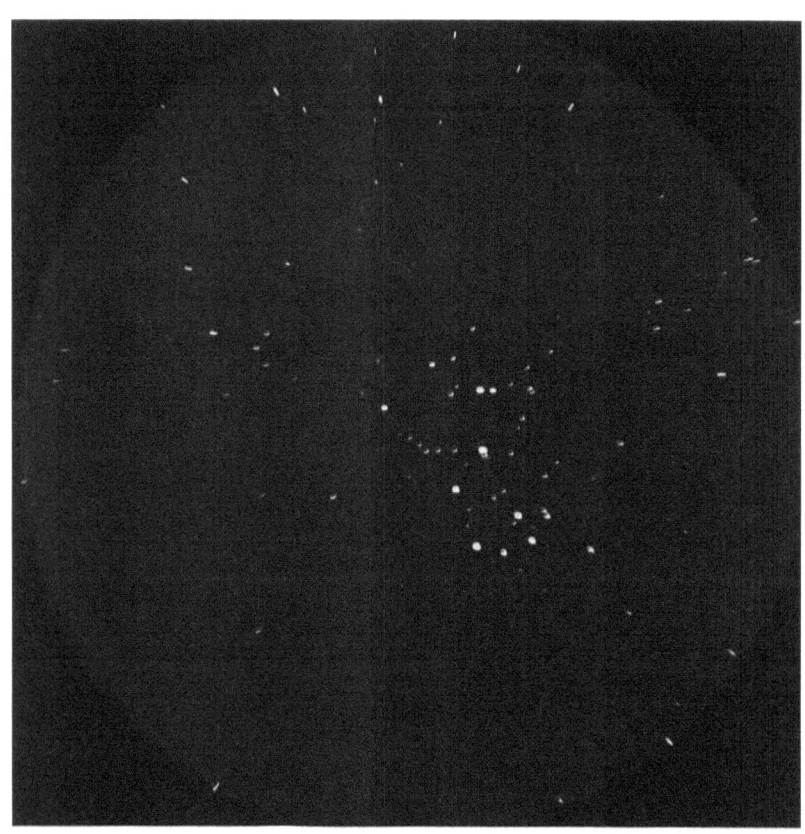

Fig 5: Pleiades (MW)

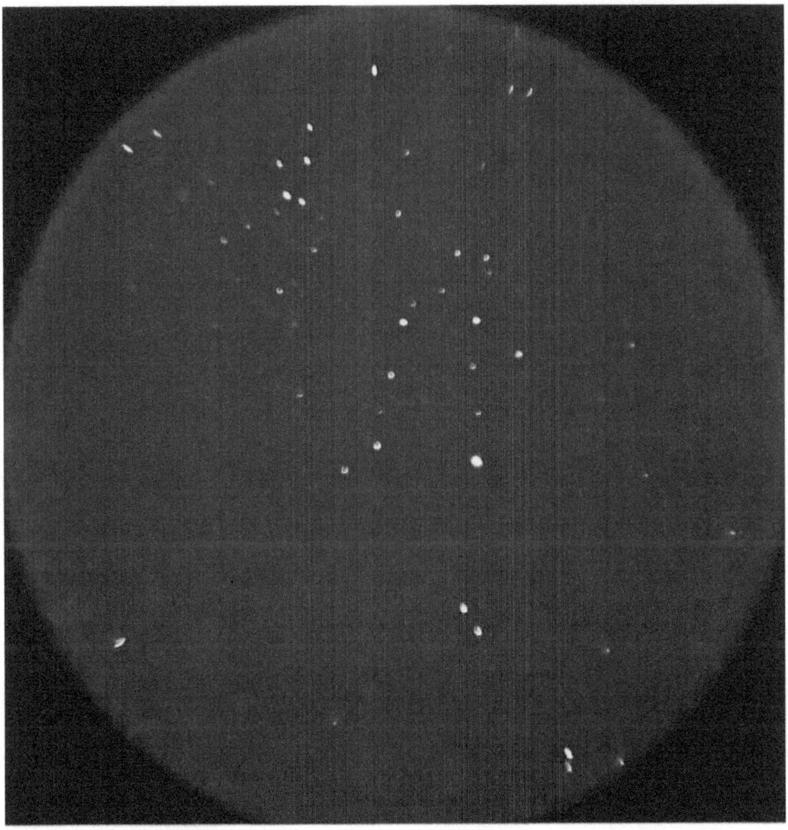

Fig 6: Perseus Cluster (MW)

Fig 7: The Moon (DFW)

Fig 8: Comma Berenices Cluster (DFW)

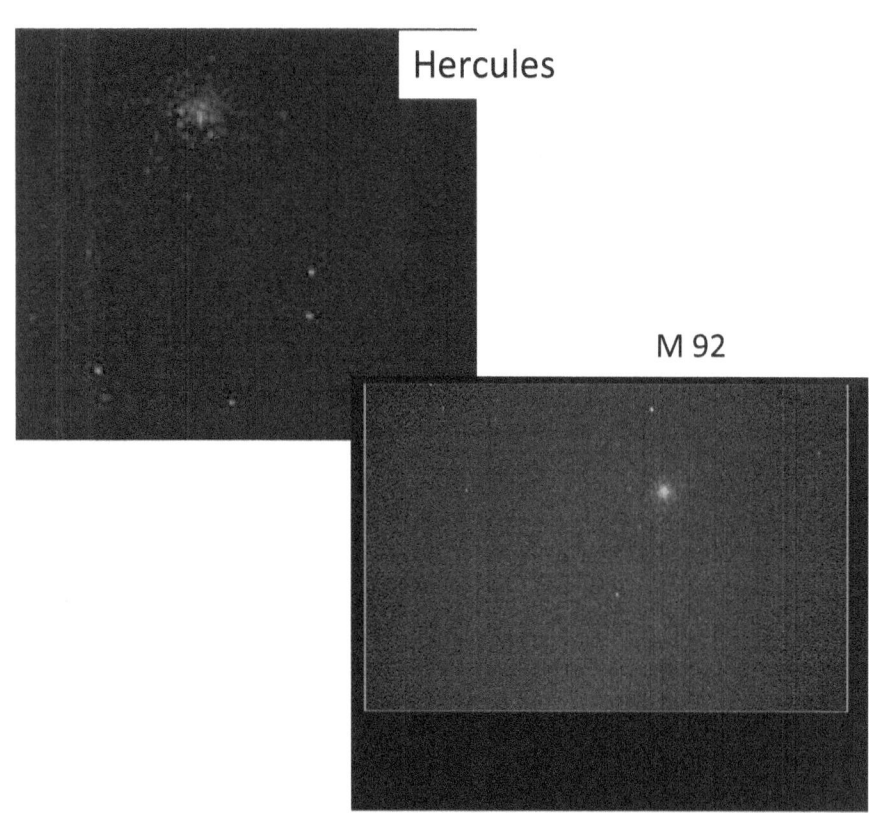

Fig 9: Hercules and M 92 Globular Clusters (DFW)

Fig 10: Alkalurops Double Star (DFW)

Fig 11: Mizar and Alcor (DFW)

Fig 12: Albireo (DFW)

Fig 13: M7 Star Cluster (DFW)

Fig 14: Cor Caroli Binary Star (DFW)

Fig 15: Hyades (DFW)

Fig 16: Moon (DFW)

Fig 17: Nu Draconis

Fig 18: Star Gamma Cygni in the center (DFW)

Fig 19: Geminin, Orion, and Auriga (MW)

Fig 20: Leo in the Center (MW)

Jucrilam saw the excitement of Zeno contemplating the pictures of the stars, and he then asked:

Jucrilam: Here in this beautiful day in Greece, feel the heat of the sun on your face and slow-moving wind, and look at this great sea.

Zeno: Reading a book of wisdom in the country of great philosophers with no duty for the next day other than philosophy and contemplation. That is what I call life in paradise.

Jucrilam: What comes to your mind when you see these worlds throughout space? What are your thoughts on these 20 pictures?

Zeno: I think that in the same way that the organs of our organism is the way it is for a purpose, I believe that there must be a purpose for there to be several worlds in this universe other than for us to just look at them.

Jucrilam: So you believe that nothing in the universe is not purposeless? If it exists it must have a goal?

Zeno: Correct! All the animals on Earth fulfill a purpose, even the Moon, so why not all of these worlds in the universe?

Jucrilam: I see. Without the Sun we would have no Earth, and without the Moon possibly no life on the Earth since the Moon kept Earth's orbit stable and protected our planet from bombardment. So, I understand your point. There must be a purpose on all those stars above.

Zeno: Indeed. Or else it makes no sense to have them in the first place.

Jucrilam: But Zeno! That would imply that nature is not blind but must be something that sees, or else if it were blind it would create things without purpose like in an accident.

Zeno: How can nature see then?

Jucrilam: I do not know for sure. Can things be exactly the way they are by an accident. Was the Big bang an accident?

Zeno: I seriously have no idea even if accidents happen in nature or if everything fulfills a purpose, a rule, that keeps the order of the cosmos as they are.

Jucrilam: Following the same reasoning. Is it possible to have a universe without evil. Everything has a purpose like you said.

Zeno: We human beings are intelligent unlike the other animals in our planet. We can have control of the sea, the air, the temperature, the wind. We can make ourselves smarter, physically more capable, and if we see something in our planet that makes no sense and that is unfair, we can fix it with our intelligence.

Jucrilam: What do you mean with that?

Zeno: Since we live in a harsh world where animals eat other animals, and so many humans struggle to survive, we can change the rules of everything. We can even change the rules of the natural world. We could teach the lions to not prey, and the ants to be friendly to us and not sting us. We can teach the bees to produce honey but be harmless to us when we get close to them. We can teach animals to not eat other animals. We humans are intelligent for a purpose and that means that we can create and change things. We can also make all humans equal to one another, an eliminate wars, and all things that are unpleasant.

Jucrilam looked at Zeno astonished.

Jucrilam: Like re-creating life on Earth! Like re-creating the human being! Like re-creating everything but with one difference! No place for evil in it! Is that then the purpose for intelligent life?

Zeno: Since everything in the universe must have a purpose, what should be the purpose of intelligent life?

Jucrilam: To have animals possessed with reason and ability to…..

Zeno: Let me finish: Ability to destroy evil and create a living world with no evil it. We can interfere with the DNA of all things, and genetic engineer a form of life with no evil it. A new lion, a new ant, a new bee, a new human.

Jucrilam: A new human that will possibly be able to live for thousands of years without death, with no disease, and physically stronger?

Zeno: Exactly! If all things in nature have a purpose, then the reason we have intelligence could be to help the universe evolve to something better than what it actually is right now!

Jucrilam: True! Everything in the universe is evolution, and what comes after intelligent life?

Zeno: Simply the end of evil is the end result of intelligent life.

Jucrilam looked at Zeno with even greater astonishment. He extended his hand and gave Zeno a handshake and said:

Jucrilam: I do not even know if that is allowed. For humans to change nature that much. It sure is something worth to think about and it could lead to success and hopefully not a failure.

Zeno: As long as we know who to give this knowledge to prevent the creation of monsters and make the Earth even more evil. There must be a law to guarantee that the changes we make will be for the good and not for bad or for the destruction.

Jucrilam: Indeed, that is something to worry about.

Zeno: And those stars you see above us in all directions in space. After having fixed all the things that are not fair in the world, and after building a human civilization where all nations prosper and where all humans live a glorious life, then we can move our beings to other worlds and bring these deadly worlds to life.

Jucrilam: In other worlds: Humans will create life in those planets and adjust the rules so that everything will be under justice.

Zeno: Yes, you got it!

Jucrilam: Welcome home Zeno. Looks like that you have reached an enlightenment of ideas here in Greece after reading from this scroll. The future has opened to you as I expected.

Zeno: Yes, I think so. I have just now eaten of the fruit of the knowledge of Good and Wisdom.

Jucrilam: Which is the world where evil makes no sense.

Zeno finished by saying: "Where evil ends in the presence of intelligence life. The re-creation of a New Heaven and a New Earth that is perfect for our eyes. All these stars are ours to bring life to them and spread the seed of life, goodness, and wisdom through all of them for all eternity.

Jucrilam: You have my signature at the bottom and all of my support.

Immediately a huge water wave appeared on the sea heading towards the shore where the two were. Zeno became desperate. From this huge wave a Xynwheel Tunnel was formed, the portal to a wormhole with the image of the face of Vloudel in it. He opened his large mouth and swallowed Jucrilam and Zeno who were taken to another dimension. The two crossed a colorful tunnel and Zeno could not tell anything anymore. All was very confusing. They were both transported to another spherical world that was entirely flat. There were no mountains, rivers, craters, or anything.

Jucrilam said to Zeno: "Here is your chance to show me how the world should be according to you. Please show me how you want this one to be!"

Zeno: Wow! It is a completely empty planet! The world must be like…..

When Zeno was about to star Jucrilam interrupted him, gave Zeno a staff and said:

Jucrilam: May it be!

There was then a huge explosion, and Zeno opened his eyes. He was in bed at home waking up from the ring of his alarm in order to get to school in the morning.

About the Author

I, Diogo Franklin de Souza, was born in the city of Rio de Janeiro, Brazil in August 20, 1986. I moved to Dallas, Texas when I was 11 years old. I write stories since I was 9 years old. My books tend to contain short summaries of the most important things I find about life, morality, religion, philosophy, and science. Like I say, everything is part of a whole system, and this is also for everything I do and write. I always wanted to have all the most important knowledge in only a few short books. That is why I write, and that is my inspiration for short summaries. I hope this book brings some inspiration also for the readers, because that really is the purpose of my work. Read it and take from it, pieces of gold for you that can be useful in your life. Enjoy….

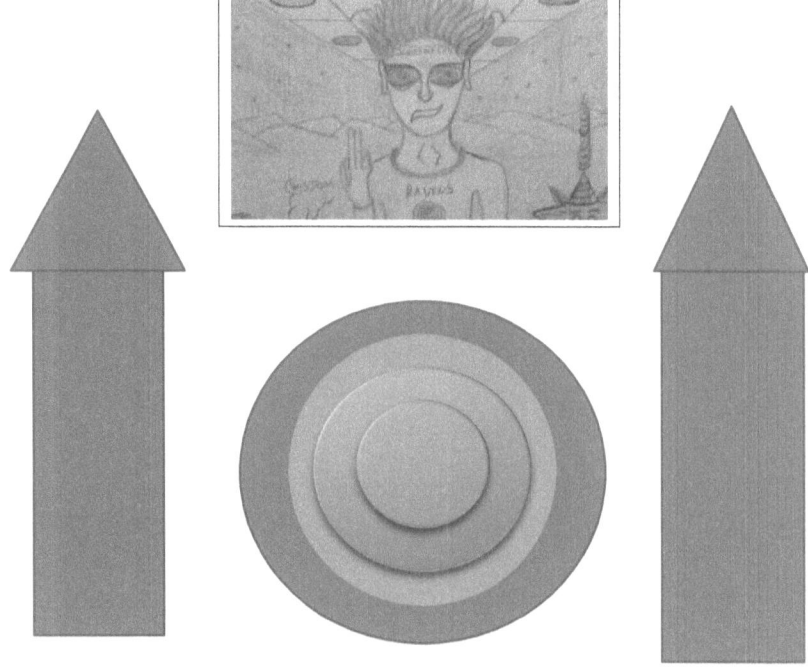

www.ingramcontent.com/pod-product-compliance
Lightning Source LLC
Chambersburg PA
CBHW030609220526
45463CB00004B/1230